SCHAUM'S *Easy* OUTLINES

INTERMEDIATE ALGEBRA

Other Books in Schaum's Easy Outlines Series Include:

Schaum's Easy Outline: Calculus
Schaum's Easy Outline: College Algebra
Schaum's Easy Outline: College Mathematics
Schaum's Easy Outline: Differential Equations
Schaum's Easy Outline: Discrete Mathematics
Schaum's Easy Outline: Elementary Algebra
Schaum's Easy Outline: Geometry
Schaum's Easy Outline: Linear Algebra
Schaum's Easy Outline: Mathematical Handbook of Formulas and Tables
Schaum's Easy Outline: Precalculus
Schaum's Easy Outline: Probability and Statistics
Schaum's Easy Outline: Statistics
Schaum's Easy Outline: Trigonometry
Schaum's Easy Outline: Bookkeeping and Accounting
Schaum's Easy Outline: Business Statistics
Schaum's Easy Outline: Economics
Schaum's Easy Outline: Principles of Accounting
Schaum's Easy Outline: Beginning Chemistry
Schaum's Easy Outline: Biology
Schaum's Easy Outline: Biochemistry
Schaum's Easy Outline: College Chemistry
Schaum's Easy Outline: Genetics
Schaum's Easy Outline: Human Anatomy and Physiology
Schaum's Easy Outline: Molecular and Cell Biology
Schaum's Easy Outline: Organic Chemistry
Schaum's Easy Outline: Applied Physics
Schaum's Easy Outline: Physics
Schaum's Easy Outline: HTML
Schaum's Easy Outline: Programming with C++
Schaum's Easy Outline: Programming with Java
Schaum's Easy Outline: Basic Electricity
Schaum's Easy Outline: Electromagnetics
Schaum's Easy Outline: Introduction to Psychology
Schaum's Easy Outline: French
Schaum's Easy Outline: German
Schaum's Easy Outline: Spanish
Schaum's Easy Outline: Writing and Grammar

SC ... ES

INTERMEDIATE ALGEBRA

BASED ON SCHAUM'S
Outline of Theory and Problems of Intermediate Algebra
BY

RAY STEEGE, M.A.
AND
KERRY BAILEY, M.A.

ABRIDGEMENT EDITOR
GEORGE J. HADEMENOS, Ph.D.

SCHAUM'S OUTLINE SERIES

McGRAW-HILL

New York Chicago San Francisco Lisbon London Madrid
Mexico City Milan New Delhi San Juan
Seoul Singapore Sydney Toronto

The McGraw-Hill Companies

RAY STEEGE received his B.A. in mathematics from the University of Wyoming and his M.A. in mathematics from the University of Northern Colorado. He taught for 10 years at East High School in Cheyenne, Wyoming, and for 25 years at Laramie Community College in Cheyenne before his retirement in 1994.

KERRY BAILEY received his B.A. in mathematics from San Diego State University and his M.A. in mathematics from the University of Colorado. He began teaching at Laramie Community College in Cheyenne, Wyoming, in 1983, and before that he taught for 10 years at Pikes Peak Community College in Colorado Springs, Colorado.

GEORGE J. HADEMENOS has taught at the University of Dallas and done research at the University of Massachusetts Medical Center and the University of California at Los Angeles. He holds a B.S. degree from Angelo State University and both M.S. and Ph.D degrees from the University of Texas at Dallas. He is the author of several books in the *Schaum's Outline* and *Schaum's Easy Outline* series.

5 6 7 8 9 0 FGR/FGR 0 9 8 7 6

ISBN 0-07-142243-9

Contents

Chapter 1
FUNDAMENTAL CONCEPTS

IN THIS CHAPTER:

- ✔ *Definitions*
- ✔ *Axioms of Equality and Order*
- ✔ *Properties of Real Numbers*
- ✔ *Operations with Real Numbers*
- ✔ *Order of Operations*

Definitions

A *set* is a collection of objects. The collection should be well defined. That is, it must be clear that an object is either in the set or is not in the set. The objects in the set are called *elements* or *members* of the set. The members of a set may be listed or a description of the members may be given.

We list the members or describe the members within braces { }. Capital letters such as A, B, C, S, T, and U are employed to name sets. For example,

$$A = \{2, 4, 6, 8, 12\} \qquad B = \{3, 6, 9, 12\}$$

$$U = \{\text{people enrolled in algebra this semester}\}$$

The symbol used to represent the phrase "is an element of" or "is a member of" is "$\in$." Thus, we write $4 \in A$ to state that 4 is a member of set A. The symbol used to represent the phrase "is not an element of" is "$\notin$." Hence, $4 \notin B$ is written to indicate that 4 is not an element of set B.

Sets are said to be equal if they contain the same elements. Hence $\{1, 5, 9\} = \{5, 9, 1\}$.

Order is disregarded when the members of a set are listed.

Sometimes a set contains infinitely many elements. In that case, it is impossible to list all of the elements. We simply list a sufficient number of elements to establish a pattern followed by a series of dots "...". For example, the set of numbers employed in counting is called the set of *natural numbers* or the set of *counting numbers*. We write $N = \{1, 2, 3, 4, \ldots\}$ to represent that infinite set. If zero is included with the set of natural numbers, the set of whole numbers is obtained. In this case, the symbol used is $W = \{0, 1, 2, \ldots\}$.

Set B above could be described as the set of multiples of three *between* 0 and 15. Note that the term "between" does not include the numbers 0 and 15. Set-builder notation is sometimes used to define sets. We write, for example,

$$B = \{x \mid x \text{ is a multiple of three between 0 and 15}\}$$

There are occasions when a set contains no elements. This set is called the *empty* or *null* set. The symbol used to represent the empty set is "$\varnothing$" or "{ }." Note that no braces are used when we represent the empty set by $\varnothing$. {negative natural numbers} is an example of an empty set.

Definition 1. Set A is a *subset* of B if all elements of A are elements of B. We write $A \subseteq B$.

Hence, if $A = \{2, 4, 6\}$ and $B = \{1, 2, 3, 4, 5, 6, 7\}$, $A \subset B$. A is called a *proper subset* of B. If $C = \{4, 2, 6\}$, $A \subseteq C$ since the sets are the same set. A is called an *improper subset* of C.

New sets may be formed by performing operations on existing sets. The operations used are union and intersection.

Definition 2. The *union* of two sets A and B, written $A \cup B$, is the set containing all of the elements in set A or B, or in both A and B.

If set-builder notation is used, we write $A \cup B = \{x | x \in A \text{ or } x \in B\}$. If $A = \{1, 5, x, z\}$ and $B = \{3, 5, 7, z\}$, then $A \cup B = \{1, 5, x, z, 3, 7\}$. Recall that the members of a set may be listed without regard to order.

Definition 3. The *intersection* of two sets A and B, written $A \cap B$, is the set containing the elements common to both sets.

If set-builder notation is used, we write $A \cap B = \{x | x \in A \text{ and } x \in B\}$. Hence, if set $A = \{1, 5, x, z\}$ and $B = \{3, 5, 7, z\}$, then $A \cap B = \{5, z\}$.

Venn diagrams are sometimes used to illustrate relationships between sets. Figure 1-1 (*a*) and (*b*) shown below illustrate the concepts discussed above. The shaded regions represent the specified set.

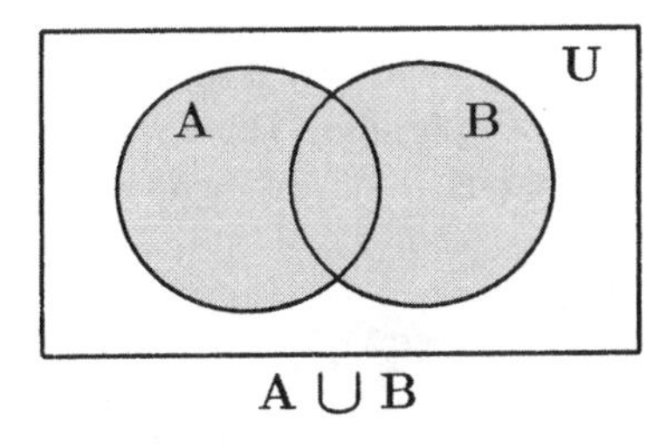

A ∪ B

Figure 1-1(*a*)

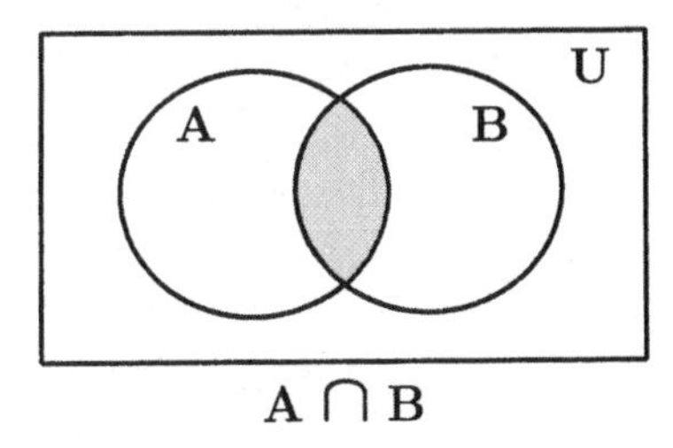

A ∩ B

Figure 1-1(*b*)

U represents the *universal* set. It is a set that contains all of the elements under discussion in a given situation. The universal set is typically represented by a rectangular region.

Figure 1-2 shows a Venn diagram that illustrates $A \subset B$. Note that all of set A is completely contained in set B.

If set S is not a subset of set T, we write $S \not\subset T$. This occurs when S contains at least one element that is not in T.

There are several additional sets of numbers that will be referred to often. Their definitions follow.

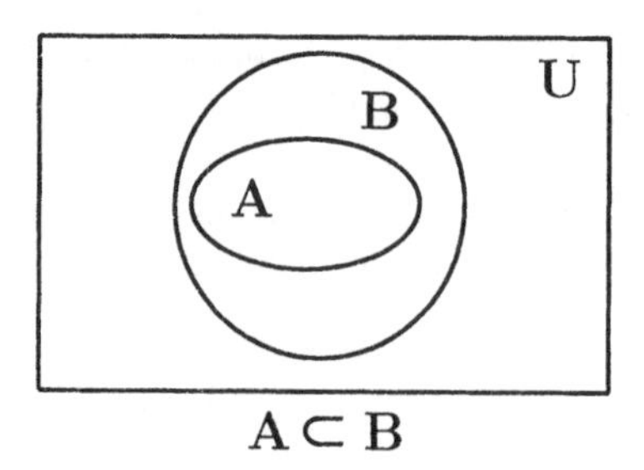

$A \subset B$

Figure 1-2

The set of *integers*, *J*, is the set of natural numbers combined with their negatives and zero. Symbolically, $J = \{\ldots, -2, -1, 0, 1, 2, \ldots\}$. Some examples are −4, −1, 0, 13, and 22.

The set of *rational numbers*, *Q*, is the set of elements which can be expressed in the form $\frac{a}{b}$, where *a* and *b* are integers and $b \neq 0$. Rational numbers can also be expressed as terminating or repeating decimals. Some examples are −2, 0, 5, $\frac{2}{3}, \frac{9}{8}$, 0.34, and −3.588.

The set of *irrational numbers* is the set of elements whose decimal representations are either nonterminating or nonrepeating. There is no single letter that is commonly used to name this set. Some examples are −1.010010001…, $\sqrt{5}$, π, and $\sqrt[3]{7}$.

The set of *real numbers*, *R*, is the set of rational numbers combined with the set of irrational numbers. That is, *R* is the union of the sets of rational and irrational numbers.

A *constant* is a symbol that represents only one number. Letters near the beginning of the alphabet, like *a*, *b*, and *c*, are typically used to represent constants. A symbol that represents a constant has only one replacement.

A *variable* is a symbol that represents any value in a specified set. Letters near the end of the alphabet, like *x*, *y*, and *z*, are typically used to represent variables. The set of possible replacements of the variable is called the *domain* of the variable. The domain of a variable is assumed to be the set of real numbers *R* if no other domain is specified.

Axioms of Equality and Order

In mathematics, we make formal assumptions about real numbers and their properties. These formal assumptions are called *axioms* or *postulates*. The terms *property*, *principle*, or *law* are sometimes used to refer to those assumptions although these terms may also refer to some consequences of those relationships we assume to be true. The assumptions can be made without restriction although their consequences should be useful.

An *equality* is a statement that symbols, or groups of symbols, represent the same quantity. The symbol used is "=", read "is equal to" or "equals."

We shall assume that the "is equal to" relationship satisfies the following properties. Assume that a, b, and c are real-valued quantities.

Reflexive property: $a = a$ (A quantity is equal to itself.)

Example 1-1: $x - 6 = x - 6$.

Symmetric property: If $a = b$, then $b = a$. (If the first quantity is equal to a second, then the second is equal to the first; an equation may be reversed.)

Example 1-2: If $5 = y + 7$, then $y + 7 = 5$.

Transitive property: If $a = b$ and $b = c$, then $a = c$. (If the first quantity is equal to a second and the second quantity is equal to a third, then the first and third quantities are equal.)

Example 1-3: If $x = y - 3$ and $y - 3 = z$, then $x = z$.

Substitution property: If $a = b$, then b may be replaced by a and vice versa. (If two quantities are equal, one may be substituted for the other.)

Example 1-4: If $x - 4 = y$ and $x = z$, then $z - 4 = y$.

We associate each real number with one and only one point on a line. The line is called a *number line*. Each number is paired with only one

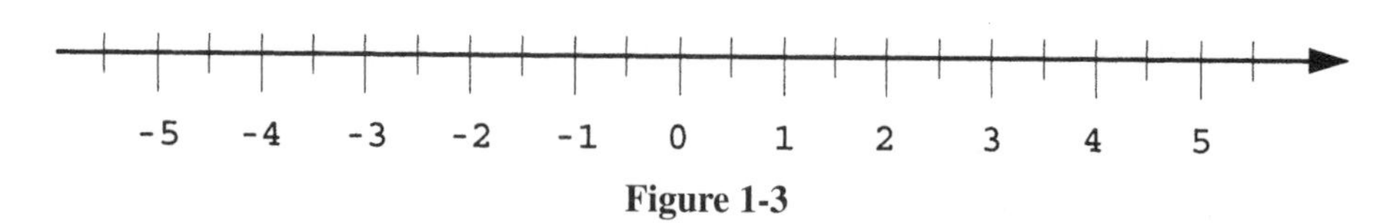

Figure 1-3

point and each point is paired with only one number. The number line is helpful in visualizing relationships between numbers. The numbers are called the *coordinates* of the points and the points are called the *graphs* of the numbers. A number line is shown in Figure 1-3. Each mark represents one-half unit. This is known as the *scale* of the drawing.

You Need to Know 

The point associated with zero is called the *origin*. Numbers associated with points to the right of the origin are *positive* while numbers associated with points to the left of the origin are *negative*. The number 0 is neither positive nor negative.

Definition 4. The number b "is less than" the number a if $a - b$ is positive.

Equivalently, the point associated with b on the number line lies to the left of the point associated with a. The symbol used to represent the "is less than" relation is "$<$." We write $2 < 4$ and $-1 < 3$ for example.

It is also true that if b is less than a, then a "is greater than" b. The symbol used to represent the "is greater than" relation is "$>$." We could express the above relationships equivalently as $4 > 2$ and $3 > -1$.

We sometimes combine the relations "is less than" and "is equal to" into one statement. In this case, the symbol used is "$\leq$." It is read "is less than or equal to." Similarly the relations of "is greater than" and "is equal to" may also be combined. The symbol used to represent this relation is

"≥." It is read "is greater than or equal to." For example, $3 \geq 0$ and $x \leq 2$ are common types of statements.

Furthermore, two conditions such as $a < b$ and $b \leq c$ are sometimes combined into the compound or double inequality $a < b \leq c$. It is read "a is less than b and b is less than or equal to c." If it is true that $a < b < c$, then b is *between* a and c. Compound or double inequalities are used to indicate "betweenness."

It is not appropriate to combine expressions such as $a < b$ and $b > c$ into the compound inequality $a < b > c$. Inequality symbols that are used in compound inequality statements must always be in the same sense. That is, they must all be "less than" relationships or "greater than" relationships.

Remember

A number that is zero or positive is said to be *nonnegative*. If $a \geq 0$, then a is nonnegative. Also, a number that is zero or negative is said to be *nonpositive*. Hence, if $a \leq 0$, than a is nonpositive. For example, 6 and 0 are nonnegative while –3 and 0 are nonpositive.

The following property of order of real numbers is assumed to be valid.

Transitive property: If $a < b$ and $b < c$, then $a < c$. (If the first quantity is less than a second and the second is less than the third, then the first is less than the third.)

Example 1-5: If $x < y$ and $y < 7$, then $x < 7$.

Subsets of real numbers are sometimes represented using interval notation. For example $\{x | a < x \leq b\}$, is written as $(a, b]$. The parenthesis is used to indicate that a is not included in the interval. The bracket indicates that b is included in the interval.

Other intervals are represented as indicated below:

(a, b) represents the real numbers between a and b. It is an *open* interval. Neither endpoint is included.
$[a, b]$ represents the real numbers between a and b inclusive. It is a *closed* interval. The endpoints are included.
$[a, b)$ represents the real numbers between a and b, including a but not b. It is a *half-open* interval. The endpoint "a" is included while the endpoint "b" is not included. The interval $(a, b]$ is also half-open.

The graphs of some subsets of the real numbers extend forever in one or both directions. The intervals used to represent those subsets are represented using *infinity* "∞" or the *negative infinity* "$-\infty$" symbols. The infinity symbol does not represent a real number; it represents a concept. For that reason, it is never included in an interval. Some representative intervals are represented below:

(a, ∞) represents all real numbers greater than a
$(-\infty, b)$ represents all real numbers less than b
$(-\infty, b]$ represents all real numbers less than or equal to b
$(-\infty, \infty)$ represents all real numbers.

Properties of Real Numbers

The following eleven properties of real numbers are assumed to be true.

Closure property of addition $a + b$ is a unique real number
Closure property of multiplication ab is a unique real number
(The sum and product of two real numbers is a unique real number.)

Example 1-6:

$2 + 4 = 6$ $\quad$ $-3 + 8 = 5$
$3(4) = 12$ $\quad$ $-4(6) = -24$

Associative property of addition $(a+b)+c=a+(b+c)$
Associative property of multiplication $(ab)c = a(bc)$
(Real numbers may be grouped or associated differently when adding or multiplying without affecting the result.)

Example 1-7:

$(3+4)+7=3+(4+7)$ $(x+5)+1=x+(5+1)$
$(4x)y=4(xy)$ $(8s)(tu)=8(stu)$

Commutative property of addition $a+b=b+a$
Commutative property of multiplication $ab=ba$
(When adding or multiplying real numbers, the result is the same if the order of the addends or factors is changed.)

Example 1-8:

$3+6=6+3$ $(x+5)+1=x+(5+1)$
$2(5)=5(2)$ $r(3)=3(r)$

Identity property of addition $a+0=0+a=a$
Identity property of multiplication $a(1)=(1)a=a$
(Zero added to a number or a number added to zero is the number; a number times one or one times a number is the number. Zero is the additive identity. One is the multiplicative identity.)

Example 1-9:

$2+0=0+2=2$ $0+(x-5)=(x-5)+0=x-5$
$3(1)=1(3)=3$ $(x-9)1=1(x-9)=x-9$

Inverse property of addition $a+(-a)=(-a)+a=0$
Inverse property of multiplication $a(1/a)=(1/a)a=1,\ a\neq 0$
[There is a unique real number $(-a)$ such that when added to a results in the additive identity zero; if $a \neq 0$, there is a unique real number $(1/a)$ such that when $(1/a)$ is multiplied by a, the result is the multiplicative identity one. $(-a)$ is the opposite of a. $(1/a)$ is the reciprocal if a.]

Example 1-10:

$2+(-2)=0$ $(-t)+t=0$
$3(1/3)=1$ $(1/s)s=1,\ s\neq 0$

Distributive properties $a(b + c) = ab + ac$

$(b + c)a = ba + ca$

(A factor may be distributed over a sum from the left or the right. Alternatively, a common factor may be factored out.)

Example 1-11:

$3(x + 4) = 3x + 3(4)$ $(s - t)5 = 5s - 5t$

$6y + 8 = 2(3y + 4)$

Operations with Real Numbers

The *absolute value* of a real number is the distance between that number and zero on the number line. We write " $|q|$ " to represent the absolute value of the quantity q. The expression is read as "the absolute value of q." The definition of absolute value of q follows:

Definition 5.

$$|q| = \begin{cases} q \text{ if } q \geq 0 \\ -q \text{ if } q < 0 \end{cases}$$

In other words, the absolute value of a positive quantity is that quantity, while the absolute value of a negative quantity is the opposite of that quantity. The absolute value of zero is zero.

Note!

The opposite of a positive number is negative and the opposite of a negative number is positive.

Addition

Addition of two quantities is the operation which associates those quantities with a third quantity called the sum. We employ the plus sign, "+," to represent the operation of addition. Thus, we write $s + t$ to represent the sum of s and t.

Addition of real numbers may be thought of as movement along a number line. Adding a positive number is associated with movement to the right while adding a negative number is associated with movement to the left. We must always start at zero.

Subtraction

We define the *difference* of two quantities a and b as $a + (-b)$. In other words, the difference of a and b is the sum of a and the additive inverse (opposite) of b. The process of finding the difference of two quantities is called *subtraction*.

Multiplication

Multiplication is an operation that associates with each pair of quantities a third quantity called the *product*. We write st or $s(t)$ or $(s)(t)$ to represent the product of s and t. Each of the quantities we multiply is called a *factor*.

Division

We define the *quotient* of a and b or a/b as the product of a and $1/b$, $b \neq 0$. In other words, dividing a by b is equivalent to multiplying a by the multiplicative inverse (reciprocal) of b.

Order of Operations

If multiple operations are to be performed in an expression, we must agree which operations are to be performed first in order to avoid ambiguity. The following order must be adhered to:

1. Operations within symbols of inclusion such as parentheses, brackets, fraction bars, etc., must be performed first. Begin with the innermost group.
2. Begin by evaluating powers in any order.
3. Next, multiplications and divisions are performed in order from left to right.
4. Last, perform additions and subtractions as encountered in order from left to right.

Solved Problem 1.1. Perform the indicated operations in: (a) $2 + 3 \cdot 6$; (b) $4[8 - (-3)]$; (c) $\dfrac{-20+4}{-2-6}$

Solution: (a) $2 + 3 \cdot 6 = 2 + 18 = 20$

(b) $4[8 - (-3)] = 4[8 + 3] = 4[11] = 44$

(c) $\dfrac{-20+4}{-2-6} = \dfrac{-16}{-8} = 2$

Chapter 2
POLYNOMIALS

IN THIS CHAPTER:

- ✔ *Definitions*
- ✔ *Sums and Differences*
- ✔ *Products*
- ✔ *Factoring*
- ✔ *Division*

Definitions

An *algebraic expression*, or simply an *expression*, is any meaningful collection of numerals, variables, and signs of operations. Several examples of expressions follow:

$$3s^2 - 4t;\ 5x^2 - 2x + 8;\ (a - b)^3;\ \text{and}\ \frac{2s^2 - 4t}{3s + t}$$

In an expression of the form $P + Q + R + S + \cdots$, P, Q, R, and S are called the *terms* of the expression. Terms are separated by addition symbols, but may be separated by subtraction symbols also.

We now illustrate with several examples:

Expression	Number of terms
$4r + 5st$	2
$3x - 5y + 4w$	3
$5(4s - 6t)$	1

The last expression, $5(4s - 6t)$, contains two factors, but only one term. The second factor of the expression contains two terms, however. The first expression, $4r + 5st$, contains two terms. The first term of the expression has two factors, while the second term of the expression contains three factors.

A *coefficient* consists of any factor or group of factors of the product of the remaining factors in a term. In the expression $4r + 5st$ above, 4 is the coefficient of r in the first term while 5 is the coefficient of st in the second term. In addition, $5s$ is the coefficient of t, $5t$ is the coefficient of s, and st is the coefficient of 5 in the second term. Normally, the word coefficient refers to the numerical coefficient in a term. In an expression such as $s - t$, the coefficient of s is 1 and the coefficient of t is -1, since $s - t = s + (-t) = 1s + (-1)t$.

A *monomial* is an algebraic expression of the form cx^n, or a product of such expressions, where c is a real number and n is a whole number. Some examples of monomials are:

$$5x^2,\ 2st^3,\ w, \text{ and } 9$$

A *polynomial* is an algebraic expression whose terms are monomials. A polynomial contains a finite number of terms, and it may contain more than one variable. A *binomial* is a polynomial that contains two terms. Some examples of binomials are:

$$4x - 8;\ x^2 + y^2;\ 3r^2s - 4; \text{ and } (x + y)^3 + 27$$

A polynomial in one variable is an expression of the form

$$a_nx^n + a_{n-1}x^{n-1} + a_{n-2}x^{n-2} + \cdots + a_2x^2 + a_1x + a_0$$

where $a_n \neq 0$ and the a_i's are real numbers, n is a nonnegative integer (whole number), and x is a variable. In the definition above, the expressions to the lower right of the a's are called *subscripts*. In a polynomial that contains many terms, subscripts are used to distinguish one coefficient from another. Note that the subscripts of the a's are the same as the exponents of x in the various terms.

The degree of a monomial is the sum of the exponents of the variables it contains. The degree of a nonzero constant is zero, and the degree of zero is undefined. The degree of a polynomial, defined as the highest degree of any of its terms, is n.

When a polynomial is written with powers of the variable decreasing, it is said to be in *standard form*. The standard form of $8 - 5x^2 + 3x$ is $-5x^2 + 3x + 8$. In this illustration, $a_2 = -5$, $a_1 = 3$, $a_0 = 8$, and $n = 2$.

Solved Problem 2.1 Write each of the following in standard form and state the degree of each: (a) $5s - 4 + 8s^2$; (b) $x^4 + 7x^7 - x^2 + 2$; (c) $y - 7$.

Solution: (a) $8s^2 + 5s - 4$, degree 2
(b) $7x^7 + x^4 - x^2 + 2$, degree 7
(c) $y - 7$, degree 1

Sums and Differences

The distributive property, as stated earlier, is given as

$$a(b+c) = ab + ac$$

The symmetric property allows us to rewrite the same property in the form

$$ab + ac = a(b+c)$$

We can employ the distributive property in the latter form to simplify certain polynomials. The following examples illustrate the idea.

Example 2.1 (a) $3x + 5x = (3 + 5)x = 8x$
(b) $4t^2 + 9t^2 = (4 + 9)t^2 = 13t^2$

Terms that involve the same powers of the variables are called *like terms.* The process illustrated in the examples above is called *combining like terms.* In each case, the result represents the same real number as the original expression for all real-numbered replacements of the variable or variables. Unlike terms cannot be combined.

Expressions that represent the same real number for all replacements of the variable(s) are called *equivalent expressions.* Hence, equivalent expressions were obtained when we simplified the expressions in the above examples.

If two expressions represent different real numbers for some combination of replacements for the variable(s) involved, the expressions are not equivalent expressions. For example, $-(s + t)$ and $-s + t$ are not equivalent expressions. If we replace s by 4 and t by 3, we find that $-(s + t) = -(4 + 3) = -7$, but $-s + t = -4 + 3 = -1$.

You Need to Know 

It is useful to think of the signs in an expression as being signs of the coefficients, such as $a - b = a + (-b)$, and the operation as being addition.

Products

Recall that the exponent tells us how many times to use the base as a factor in an exponential expression. For instance, 4^3 means $4 \cdot 4 \cdot 4$. In general, $b^m b^n$ means m factors of b times n factors of b. There are $m + n$ factors of b, which may be written as b^{m+n}. This result is stated formally as an important law of exponents.

Law 1: For all positive integer exponents, $b^m b^n = b^{m+n}$.

Remember

The commutative and associative laws may be employed along with our first law of exponents to multiply monomials.

Solved Problem 2.2 Multiply (a) x^3x^2 and (b) $(2x^2)(4x)$.

Solution: (a) $x^3x^2 = x^{3+2} = x^5$

(b) $(2x^2)(4x) = (2 \cdot 4)(x^2 \cdot x) = 8x^3$

Law 2: For all positive integer exponents, $(b^m)^n = b^{mn}$.

Law 3: For all positive integer exponents, $(ab)^n = a^n b^n$.

The second law merely states that if some power of an expression is raised to a power, the result is that expression raised to the product of those powers. Similarly, the third law states that a power of a product of two expressions is the product of the same power of the expression.

We can generalize the distributive property using the associative property as shown below:

$$\begin{aligned} a(b+c+d) &= a[(b+c)+d] \\ &= a(b+c)+ad \\ &= ab+ac+ad \end{aligned}$$

Observe that the factor a is multiplied by each term within the parentheses.

The same properties can be applied to find products of polynomials containing more than one term.

$$\begin{aligned}(a+b)(c+d) &= (a+b)c+(a+b)d \\ &= ac+bc+ad+bd\end{aligned}$$

The result simply says that every term in the first factor is multiplied by every term in the second factor.

Special Products

$$(a+b)^2 = (a+b)(a+b) = a^2+ab+ab+b^2 = a^2+2ab+b^2$$

$$(a-b)^2 = (a-b)(a-b) = a^2-ab-ab+b^2 = a^2-2ab+b^2$$

$$(a+b)(a-b) = a^2-ab+ab-b^2 = a^2-b^2$$

$$(a+b)^3 = a^3+3a^2b+3ab^2+b^3$$

$$(a-b)^3 = a^3-3a^2b+3ab^2-b^3$$

Factoring

The first type of factoring we shall consider is that of factoring expressions that contain common monomial factors in their terms. Recall that monomials consist of constants or a product of a constant and one or more variables raised to positive integer powers.

The first step entails identifying factors that occur in every term of the expression. These common factors will be factored out of each term, thus obtaining a product of factors. We say that the expression is factored completely when there are no common factors remaining in the terms other than the number one. (One is a factor of every expression.) An expression that is factored is said to be *prime*.

Solved Problem 2.3 Factor completely (a) $4s + 4t$ and (b) $5x^2y - 50z$.

Solution: (a) $4s + 4t = 4(s + t)$
(b) $5x^2y - 50z = 5(x^2y - 10z)$

Factoring by Grouping

The concept of a common monomial factor can be extended to common binomial factors. Some polynomials with an even number of terms can frequently be factored by grouping terms with a common factor. For example, the polynomial $ax + bx + 3a + 3b$ or $(ax + bx) + (3a + 3b)$ may be written as $(a + b)x + (a + b)^3$ using the distributive law twice: once for the first pair of factors and once for the last pair. Now observe that the two groups contain the common factor $(a + b)$. This common factor can be factored out to obtain $(a + b)(x + 3)$. This, the factored form of the original expression has been obtained.

There are some general guidelines we should adhere to when factoring expressions in general. They are the following:

1. Identify and remove, that is, factor out all common monomial factors first.

2. If the expression contains two terms, it may be the difference of two squares, or a sum or difference of cube terms. If so, apply the appropriate pattern. Recall that the sum of two squares is prime.

3. If the expression contains three terms, determine if two of those terms are perfect squares. If that is the case, the expression may be a perfect square binomial. Otherwise, it may be a general form.

4. If the expression contains four terms, determine if two of those terms are perfect cubes. If that is the case, the expression may be a perfect cube binomial. Otherwise, it may factor by grouping.

Solved Problem 2.4 Factor completely (a) $xs + xt + 3s + 3t$ and (b) $2x^2 + 3x + 4x + 6$.

Solution:

(a) $$\begin{aligned} xs + xt + 3s + 3t &= (xs + xt) + (3s + 3t) \\ &= x(s + t) + 3(s + t) \\ &= (x + 3)(s + t) \end{aligned}$$

(b) $$\begin{aligned} 2x^2 + 3x + 4x + 6 &= \left(2x^2 + 3x\right) + (4x + 6) \\ &= x(2x + 3) + 2(2x + 3) \\ &= (x + 2)(2x + 3) \end{aligned}$$

Division

Recall that b^n means that b is used as a factor n times when n is a positive integer. Consider the expression b^m/b^n.

If the exponents m and n are positive integers, the numerator represents m factors of b, and the denominator represents n factors of b. Therefore, if $b \neq 0$ and $m > n$,

$$\begin{aligned}
\frac{b^m}{b^n} &= \frac{\overbrace{b \cdot b \cdot b \cdot \cdots \cdot b}^{m \text{ factors}}}{\underbrace{b \cdot b \cdot b \cdot \cdots \cdot b}_{n \text{ factors}}} \\
&= \frac{\left(\overbrace{b \cdot b \cdot b \cdot \cdots \cdot b}^{n \text{ factors}}\right)\left(\overbrace{b \cdot b \cdot b \cdot \cdots \cdot b}^{m-n \text{ factors}}\right)}{\underbrace{b \cdot b \cdot b \cdot \ldots \cdot b}_{n \text{ factors}}} \\
&= \left(\underbrace{1 \cdot 1 \cdot 1 \cdot \cdots \cdot 1}_{n \text{ factors}}\right)\left(\underbrace{b \cdot b \cdot b \cdot \cdots \cdot b}_{m-n \text{ factors}}\right) \\
&= b^{m-n}
\end{aligned}$$

The last lines follow since $b/b = 1$ and $1 \cdot b = b$.

Similarly, if $n > m$,

$$\begin{aligned}
\frac{b^m}{b^n} &= \frac{\overbrace{b \cdot b \cdot b \cdot \cdots \cdot b}^{m \text{ factors}}}{\underbrace{b \cdot b \cdot b \cdot \cdots \cdot b}_{n \text{ factors}}} = \frac{\overbrace{b \cdot b \cdot b \cdot \cdots \cdot b}^{m \text{ factors}}}{\left(\underbrace{b \cdot b \cdot b \cdot \cdots \cdot b}_{m \text{ factors}}\right)\left(\underbrace{b \cdot b \cdot b \cdot \cdots \cdot b}_{n-m \text{ factors}}\right)} \\
&= \frac{1}{\left(\underbrace{b \cdot b \cdot b \cdot \cdots \cdot b}_{n-m \text{ factors}}\right)} = \frac{1}{b^{n-m}}
\end{aligned}$$

We are merely dividing out common factors that occur in both numerator and denominator in the above illustrations. The result in both instances is simply b raised to the positive difference of the powers m and n written in the numerator and denominator, whichever results in a positive exponent on b.

We'll now consider $(a/b)^n$ where $b \neq 0$ and n is a positive integer. We observe that

$$\left(\frac{a}{b}\right)^n = \overbrace{\left(\frac{a}{b}\right)\left(\frac{a}{b}\right)\left(\frac{a}{b}\right)\cdot \cdots \cdot\left(\frac{a}{b}\right)}^{n \text{ factors}} = \frac{\overbrace{a\cdot a\cdot a\cdot \cdots \cdot a}^{n \text{ factors}}}{\underbrace{b\cdot b\cdot b\cdot \cdots \cdot b}_{n \text{ factors}}} = \frac{a^n}{b^n}.$$

We have shown that if we are raising a quotient to some power n, the result is the numerator raised to that power divided by the denominator raised to the same power.

The results of all of the laws of exponents we have discussed are now restated for your reference. They are employed in expressions involving positive, integer exponents.

Laws of Exponents

1. $b^m \cdot b^n = b^{m+n}$
2. $\left(b^m\right)^n = b^{m \cdot n}$
3. $(ab)^n = a^n b^n$

4a. $\dfrac{b^m}{b^n} = b^{m-n}$ if $b \neq 0$ and $m > n$

4b. $\dfrac{b^m}{b^n} = \dfrac{1}{b^{n-m}}$ if $b \neq 0$ and $n > m$

5. $\left(\dfrac{a}{b}\right)^n = \dfrac{a^n}{b^n}$ if $b \neq 0$

Solved Problem 2.5 Divide and simplify (a) $\dfrac{x^6}{x^4}$, (b) $\dfrac{3x^3}{12x^5}$, (c) $\dfrac{10x^4+20x^3}{5x^2}$, and (d) $\dfrac{5x^3y-7x^2y^2+12xy^3}{xy}$

Solution:

(a) $\dfrac{x^6}{x^4} = x^{6-4} = x^2$

(b) $\dfrac{3x^3}{12x^5} = \dfrac{1}{4x^{5-3}} = \dfrac{1}{4x^2}$

(c) $\dfrac{10x^4 + 20x^3}{5x^2} = \dfrac{10x^4}{5x^2} + \dfrac{20x^3}{5x^2} = 2x^2 + 4x$

(d) $\dfrac{5x^3y - 7x^2y^2 + 12xy^3}{xy} = \dfrac{5x^3y}{xy} - \dfrac{7x^2y^2}{xy} + \dfrac{12xy^3}{xy} = 5x^2 - 7xy + 12y^2$

Chapter 3
Rational Expressions

In This Chapter:

- ✔ *Basic Properties*
- ✔ *Products and Quotients*
- ✔ *Sums and Differences*
- ✔ *Mixed Operations and Complex Fractions*

Basic Properties

Recall that a rational number has the form a/b, where a and b are integers and $b \neq 0$.

Similarly, a rational expression is the quotient (ratio) of two polynomials. It is an algebraic fraction defined for all real values of the variable(s) in the numerator and denominator provided the denominator is not equal to zero. Be reminded that polynomials are expressions whose variables have nonnegative integer exponents.

Any particular quotient (fraction) can be written in infinitely many forms. For example,

$$\frac{1}{2} = \frac{2}{4} = \frac{3}{6} = \cdots \quad \text{and} \quad \frac{2}{7} = \frac{6}{21} = \frac{18}{63} = \cdots$$

In each case, we obtain an equivalent fraction by multiplying the numerator and denominator by the same nonzero constant, say k. Equivalent expressions result since we are actually multiplying the fraction by one, $(k/k = 1)$, and one is the multiplicative identity element.

The principle we have been illustrating is called the *fundamental principle of fractions*. Symbolically, we write the following:

Fundamental Principle of Fractions

$$\frac{a}{b} = \frac{a}{b} \cdot 1 = \frac{a}{b} \cdot \frac{k}{k} = \frac{ak}{bk} \quad \text{where } b, k \neq 0$$

In the above expression, a, b, and k each represent a polynomial. Some or all may also be constants, since constants are zero-degree polynomials. We now may divide out the same factor(s) that occur in the numerator and denominator. Recall that factors are quantities that are multiplied, while terms are quantities that are added or subtracted. Equivalent expressions are not obtained when *terms* are divided out. DO NOT DIVIDE OUT (CANCEL) TERMS!

Solved Problem 3.1 Reduce the following quotients to lowest terms:

(a) $\frac{20}{42}$, (b) $\frac{4x+6}{14}$, (c) $\frac{(p+t)^2}{2p+2t}$, and (d) $\frac{x^2 - y^2}{3x^2 + 5xy + 2y^2}$

Assume no denominator is zero.

Solution:

(a) $\frac{20}{42} = \frac{2 \cdot 10}{2 \cdot 21} = \frac{10}{21}$

(b) $\frac{4x+6}{14} = \frac{2(2x+3)}{2 \cdot 7} = \frac{2x+3}{7}$

(c) $\frac{(p+t)^2}{2p+2t} = \frac{(p+t)(p+t)}{2(p+t)} = \frac{p+t}{2}$

(d) $\frac{x^2 - y^2}{3x^2 + 5xy + 2y^2} = \frac{(x+y)(x-y)}{(x+y)(3x+2y)} = \frac{x-y}{3x+2y}$

Products and Quotients

We now possess the essential tools for performing operations on rational expressions. We shall begin with the operation of multiplication since it is the simplest. The definition of multiplying rational expressions is identical for arithmetic and algebraic quantities.

Definition of Multiplication

$$\frac{a}{b}\cdot\frac{c}{d}=\frac{ac}{bd}, \text{ where } b,d\neq 0$$

In words, the product of two fractions is the product of the numerators divided by the product of the denominators, provided neither denominator is zero. If there are common factors in any of the numerators and denominators, they should be divided out prior to multiplying the fractions.

Solved Problem 3.2 Write each product as a single fraction in lowest terms: (a) $\frac{3}{16}\cdot\frac{4}{15}$ and (b) $\frac{5a^2}{b}\cdot\frac{b^3}{10a}$

Solution:

(a) $\frac{3}{16}\cdot\frac{4}{15}=\frac{1}{4}\cdot\frac{1}{5}=\frac{1}{20}$

(b) $\frac{5a^2}{b}\cdot\frac{b^3}{10a}=\frac{ab^2}{2}$

Our next objective is to learn how to divide rational expressions. We know that

$$18\div 6=\frac{18}{6}=3 \text{ since } 3\cdot 6=18$$

That is, quotient times divisor is equal to the dividend. We have also learned that the reciprocal of a quantity is one divided by the quantity. Therefore,

$$\frac{1}{\frac{c}{d}}=1\div\frac{c}{d}=1\cdot\frac{d}{c}=\frac{d}{c}$$

In addition, the definition of multiplication tells us that

$$\frac{ad}{bc} = \frac{a}{b} \cdot \frac{d}{c}$$

We employ this equivalence to formally state the definition of division of rational expressions:

Definition of Division

$$\frac{a}{b} \div \frac{c}{d} = \frac{a}{b} \cdot \frac{d}{c}, \quad \text{where } b, c, d \neq 0$$

In words, the quotient of two fractions is obtained by multiplying the dividend by the multiplicative inverse (reciprocal) of the divisor. The operation of division is the inverse of multiplication, just as subtraction is the inverse of addition. Inverse operations undo each other.

Solved Problem 3.3 Divide and simplify by reducing to lowest terms: (a) $\frac{3}{4} \div \frac{25}{16}$ and (b) $\frac{5}{8} \div 3$

Solution:

(a) $\frac{3}{4} \div \frac{25}{16} = \frac{3}{4} \cdot \frac{16}{25} = \frac{12}{25}$

(b) $\frac{5}{8} \div 3 = \frac{5}{8} \cdot \frac{1}{3} = \frac{5}{24}$

Sums and Differences

Same Denominator

We employ the definition of division and the distributive property to obtain expressions that demonstrate how to add and subtract rational expressions. Hence, if $c \neq 0$,

$$\frac{a}{c}+\frac{b}{c}=a\left(\frac{1}{c}\right)+b\left(\frac{1}{c}\right) \quad \text{Definition of division}$$

$$=(a+b)\left(\frac{1}{c}\right) \quad \text{Distributive property}$$

$$=\frac{a+b}{c} \quad \text{Definition of division}$$

The above sequence of steps defines how fractions are added. We now employ that result and the definition of subtraction of real numbers (i.e., $a-b=a+(-b)$) to define subtraction of fractions.

$$\frac{a}{c}-\frac{b}{c}=\frac{a}{c}+\left(\frac{-b}{c}\right) \quad \text{Definition of subtraction}$$

$$=\frac{a+(-b)}{c} \quad \text{Definition of addition}$$

$$=\frac{a-b}{c} \quad \text{Definition of subtraction}$$

We now state the definitions of addition and subtraction of rational expressions formally.

Definition of Addition

$$\frac{a}{c}+\frac{b}{c}=\frac{a+b}{c}, c\neq 0$$

Definition of Subtraction

$$\frac{a}{c}-\frac{b}{c}=\frac{a-b}{c}, c\neq 0$$

In words, to add or subtract fractions that contain the same denominator, we merely add or subtract, respectively, the numerators and divide the result by the denominator.

Solved Problem 3.4 Perform the indicated operation: (a) $\frac{2}{7}+\frac{4}{7}$ and (b) $\frac{x}{5}+\frac{y}{5}$

Solution:

(a) $\frac{2}{7}+\frac{4}{7}=\frac{2+4}{7}=\frac{6}{7}$

(b) $\frac{x}{5}+\frac{y}{5}=\frac{x+y}{5}$

Unlike Denominator

If the fractions involved possess different denominators, we must rewrite the fractions as equivalent fractions that have the same denominator. We employ the fundamental principle of fractions

$$\frac{a}{b}=\frac{ak}{bk},\ b,k \neq 0,$$

to accomplish the task.

The primary task when finding equivalent fractions is that of determining the appropriate k.

The denominator bk must be an expression that is exactly divisible by each of the original denominators. The expression desired for bk is the "smallest" quantity that each of the original denominators divides exactly. It is called the *least common denominator* or *LCD* of the fractions.

The expression that k represents is found by first factoring all of the denominators into prime factors. Next divide the factors of each individual denominator into the LCD. The quotient is the appropriate expression to use for k in each case. The k expression is the product of all of the factors of the denominator bk = LCD which are not common to the LCD.

You Need to Know 

To find the LCD of two or more fractions:

1. Find the prime factors of each denominator.
2. The LCD is the product of the different factors that occur in the various denominators. Repeated factors are used the largest number of times they occur in any particular denominator.

Solved Problem 3.5 Find the LCD for the following fractions:

$$\frac{6}{175}, \frac{5}{36}, \text{ and } \frac{8}{135}.$$

Solution:

$175 = 5^2 \cdot 7; 36 = 2^2 \cdot 3^2$; and $135 = 5 \cdot 3^3$. The LCD = 18,900.

Now that we can rewrite fractions with different denominators as equivalent fractions having the same denominator, we are in a position to find sums and differences of all fractions. The process is summarized below.

1. Find the LCD.
2. Rewrite each fraction as an equivalent fraction with the LCD as its denominator.
3. Apply the definition of addition or subtraction of fractions with the same denominator.

Mixed Operations and Complex Fractions

We sometimes are required to perform some combination of the four fundamental operations on rational expressions or fractions. We now possess the tools needed for the task.

Solved Problem 3.6 Perform the indicated operations and simplify:

$$\left(\frac{2}{x}+\frac{1}{y}\right) \div \frac{1}{x^2y}.$$

Solution:

$$\begin{aligned}
\left(\frac{2}{x}+\frac{1}{y}\right) \div \frac{1}{x^2y} &= \left(\frac{2}{x}\cdot\frac{y}{y}+\frac{1}{y}\cdot\frac{x}{x}\right) \div \frac{1}{x^2y} \\
&= \frac{2y+x}{xy} \div \frac{1}{x^2y} \\
&= \frac{2y+x}{xy} \cdot \frac{x^2y}{1} \\
&= (2y+x)x \\
&= x^2 + 2xy
\end{aligned}$$

Chapter 4
First-Degree Equations and Inequalities

In This Chapter:

- ✔ *Solving First-Degree Equations*
- ✔ *Graphs of First-Degree Equations*
- ✔ *Formulas and Literal Equations*
- ✔ *Applications*
- ✔ *Solving First-Degree Inequalities*
- ✔ *Graphs of First-Degree Inequalities*
- ✔ *Applications Involving Inequalities*
- ✔ *Absolute-Value Equations and Inequalities*

Solving First-Degree Equations

Solving various kinds of equations and inequalities is one of the most essential and useful skills in mathematics. In this chapter, we will develop techniques for solving first-degree equations and inequalities. We shall assume that variables may be replaced by any *real number*.

Definition 1. An *equation* is a mathematical statement that two expressions are equal.

Equations may be true always, sometimes, or never.

Definition 2. An equation that is always true is called an *identity*.

Some examples of identities are $x + 2 = 2 + x$ and $x^2 = (-x)^2$. Each is true for every real value of x.

Definition 3. An equation that is sometimes true is called a *conditional* equation.

Some examples of conditional equations are $x + 2 = 8$ and $x^2 - x - 6 = 0$. Each is true for specific values of x, 6 for the first equation and both 3 and –2 for the second, but false for all other values. The equations are true for certain "conditions" only.

Definition 4. An equation that is never true is called a *contradiction*.

Some examples of contradictions are $x + 2 = x$ and $3 = 5$. Neither is true for any value of x.

Definition 5. A value of the variable that results in a true equation is called a *solution* or *root* of the equation. The value is said to "satisfy" the equation.

Definition 6. The set of all of the solutions of an equation is called the *solution set* of the equation.

Verify that $x = 8$ is a solution of each of the following equations. Replace x by 8 and observe that a true statement results in each case.

(i) $x = 8$
(ii) $3x = 24$
(iii) $3x - 3 = 21$
(iv) $3(x - 1) + 5 = 26$

Definition 7. Equations that have exactly the same solution set are called *equivalent equations.*

Hence, the equations that appear immediately prior to the preceding definition are all equivalent equations.

The solution to the first equation in the sequence above is obvious. The solution to each of the successive equations in the sequence is increasingly more difficult to verify.

In general, the process of solving an equation involves transforming the equation into a sequence of equivalent equations. The sequence should result in "simpler" equations as we proceed. Ultimately we hope to obtain an equation whose solution is obvious. We employ the axioms and properties of equality in an appropriate order to accomplish the task.

Definition 8. A *first-degree equation* is an equation that can be written in the form $ax + b = 0$, where a and b are constants and $a \neq 0$.

A first-degree equation is an equation in which the variable appears to the first power only and is not a part of the denominator. Some examples of first-degree equations are $4x + 9 = 0$; $5x - 9 = 2x$; and $2(x + 3) - 4(2 - x) = x$.

Assuming the variables a, b, and q represent real number values or quantities, the following properties can be formally stated:

Addition property of equality

$$\text{If } a = b, \text{ then } a + q = b + q.$$

Multiplication property of equality

$$\text{If } a = b, \text{ then } aq = bq \text{ provided that } q \neq 0.$$

You Need to Know ✓

Equivalent equations are obtained when the properties of equality are applied to an equation. Those properties allow us to obtain equivalent equations when we add or subtract the same quantity to or from both sides, or when we multiply or divide both sides by the same nonzero quantity.

Solved Problem 4.1 Solve $2x + 5 = 11$ for x.

Solution:

$2x + 5 = 11$	
$2x + 5 - 5 = 11 - 5$	Addition (subtraction) property of equality
$2x = 6$	Combine like terms
$\frac{2x}{2} = \frac{6}{2}$	Multiplication (division) property of equality
$x = 3$	Simplify

Three is the *solution* or *root* of the given equation. The *solution set* is $\{3\}$. It is customary to state the result simply in the form $x = 3$.

The process used to solve first-degree equations not involving fractions is summarized below:

1. Use the distributive property and simplify each side of the equation if possible.
2. Use the addition property of equality to obtain an equation that contains the variable term on one side and the constant term on the other side.
3. Use the multiplication property of equality to make the coefficient of the variable one. In other words, isolate the variable.
4. Check the result in the original equation.

Graphs of First-Degree Equations

We now consider first-degree equations in two variables. A solution to an equation in two variables consists of a pair of numbers. A value for each variable is needed in order for the equation to be identified as true or false.

Consider the equation $y = 3x - 4$. If $x = 1$, then $y = 3(1) - 4 = -1$. So, if x is replaced by 1 and y is replaced by -1, the equation is satisfied. Other pairs of numbers that satisfy the equation can be found:

If $x = 0$, $y = -4$

If $x = 3$, $y = 5$

If $x = -2$, $y = -10$

It is customary to express the pairs of numbers as an *ordered pair*. We usually write the pair in alphabetical order and use parentheses to express the fact that a specific order is intended.

In the ordered pair (x, y), x is the *first component* or *abscissa*, and y is the *second component* or *ordinate*. The solution set consists of all the ordered pairs that satisfy the equation. The first component or *x-coordinate* of the ordered pair is the directed distance of the point from the vertical axis. The second component or *y-coordinate* of the ordered pair is the directed distance of the point from the horizontal axis. The components of the ordered pair that corresponds to a particular point are called the *coordinates* of the point, and the point is called the *graph* of the ordered pair. If a point lies to the right of the vertical axis, its *x-coordinate* is positive. On the other hand, the *x-coordinate* is negative if the point lies to the left of the vertical axis. Similarly, the *y-coordinate* is positive if the point lies above the horizontal axis and negative if the point lies below the horizontal axis. If a point lies on one of the axes, its directed distance from the axis is zero and the corresponding coordinate is zero. The coordinates of the origin are (0,0). A rectangular coordinate system is shown in Figure 4-1.

Solved Problem 4.2 Plot, that is, graph, the points with coordinates (4, 2), (–3, 3), (0, 5), (2, 0), and (–4, –3) on a rectangular coordinate system.

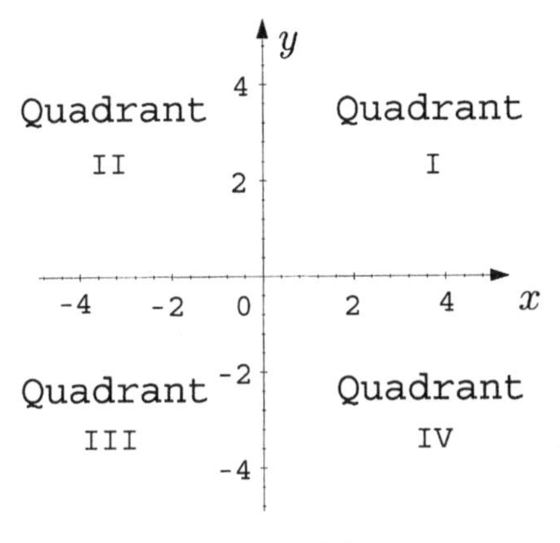

Figure 4-1

Solution: Draw the axes and label them. Then locate the points. To graph (4, 2), start at the origin and move 4 units to the right, then 2 units up. Plot the remaining points in a similar manner. See Figure 4-2.

Theorem 1. The graph of an equation of the form $ax + by = c$, where a and b are not both zero, is a straight line. Conversely, every straight line is the graph of an equation of the form $ax + by = c$.

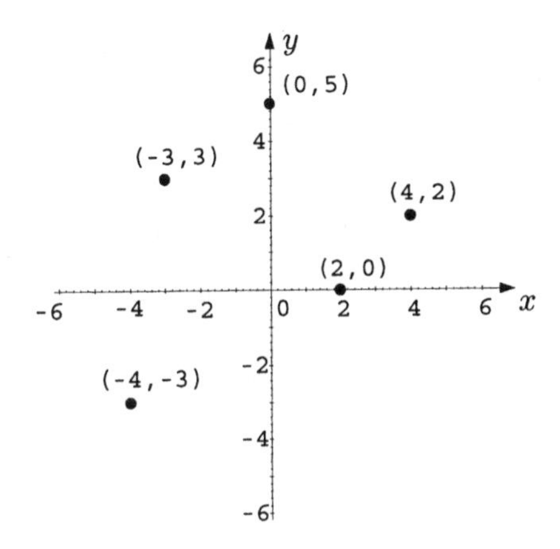

Figure 4-2

First-degree equations in the form $ax + by = c$ are called *linear equations* for this reason. The form $ax + by = c$ is commonly called the *standard form* of the equation.

Definition 9. The *x-intercepts* of a graph are the x values of the points where the graph intersects the x-axis. The *y-intercepts* of a graph are the y values of the points where the graph intersects the y-axis.

Solved Problem 4.3 Graph the equation $3x + 5y = 15$. Find the intercepts and locate a third point as a check. Label the intercept points.

Solution: Let $y = 0$ and solve for x to find the x-intercept.

$$\begin{aligned} 3x + 5y &= 15 \\ 3x + 5(0) &= 15 \\ 3x &= 15 \\ x &= 5 \text{ is the } x\text{-intercept.} \end{aligned}$$

Let $x = 0$ and solve for y to find the y-intercept.

$$\begin{aligned} 3x + 5y &= 15 \\ 3(0) + 5y &= 15 \\ 5y &= 15 \\ y &= 3 \text{ is the } y\text{-intercept.} \end{aligned}$$

Plot the points and draw the line. See Figure 4-3.

Choose $x = -5$ and solve for y to locate a third point.

$$\begin{aligned} 3x + 5y &= 15 \\ 3(-5) + 5y &= 15 \\ -15 + 5y &= 15 \\ 5y &= 30 \\ y &= 6 \end{aligned}$$

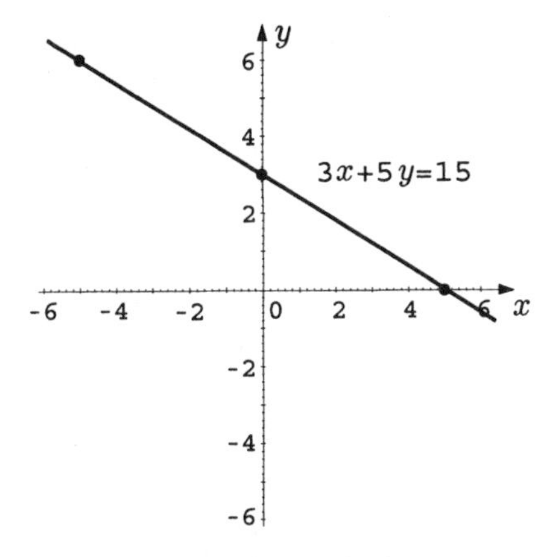

Figure 4-3

The third point has coordinates (–5, 6) and appears to lie on the line. See Figure 4-3.

Formulas and Literal Equations

We now possess the essential tools for performing operations on rational expressions. We shall begin with the operation of multiplication since it is the simplest. The definition of multiplying rational expressions is identical for arithmetic and algebraic quantities.

Definition 10. Equations that contain more than one variable are called *literal equations.*

Definition 11. Literal equations that express how quantities encountered in practical applications are related are called *formulas.*

Some examples of literal equations that are formulas are shown below:

$P = 2l + 2w$	Perimeter of a rectangle
$d = st$	Distance traveled
$I = Prt$	Simple interest

Definition 12. A literal equation is said to be *solved explicitly* for a variable if that variable is isolated on one side of the equation.

Applications

Solving real-world problems very often entails translating verbal statements into appropriate mathematical statements. Unfortunately, there is no simple method available to accomplish this.

We offer the following suggestions as an orderly approach to apply:

1. Read the problem carefully! You may need to read it several times in order to understand what is being said and what you are asked to find.
2. Draw diagrams or figures whenever possible. This will help you analyze the problem. Your figure should be drawn and labeled as accurately as possible in order to avoid wrong conclusions.
3. Identify the unknown quantity (or quantities) and use a variable to label it (them). The first letter of a key word may be a good letter to use. Write a complete sentence that states explicitly what your variable represents. Don't be ambiguous or vague.
4. Determine how the known and unknown quantities are related. The words of the problem may tell you. If not, there may be a particular formula that is relevant.
5. Write an equation that relates the known and unknown quantities. Be careful when doing this! Ask yourself if the equation translates the words of the problem accurately. Your equation must seem reasonable and make sense.
6. Solve the equation.
7. Answer the question that was asked. The value of the variable you used may not be the answer to the question. It depends upon how you defined the variable you are using.
8. Check the answer in the statement of the problem.

Definition 13. The quotient of two quantities, a/b, is called a *ratio.*

Definition 14. A statement that two ratios are equal, $a/b = c/d$, is called a *proportion.*

Solved Problem 4.4 An automobile uses 7 gallons of gasoline to travel 154 miles. How many gallons are required to make a trip of 869 miles?

Solution: Let g be the number of gallons required to make the trip. The ratio of the gallons equals the ratio of the miles. Therefore,

$$\frac{g}{7} = \frac{869}{154}$$

$$g = \frac{7(869)}{154} = 39.5$$

We conclude that 39.5 gallons will be required to make a trip of 869 miles.

Solved Problem 4.5 A rectangle is three feet longer than twice its width. If the perimeter is ninety feet, what are the dimensions of the rectangle?

Solution: We are asked to find the dimensions, that is, the length and width, of the rectangle. Now draw a rectangle and label the dimensions. Refer to Figure 4-4 below.

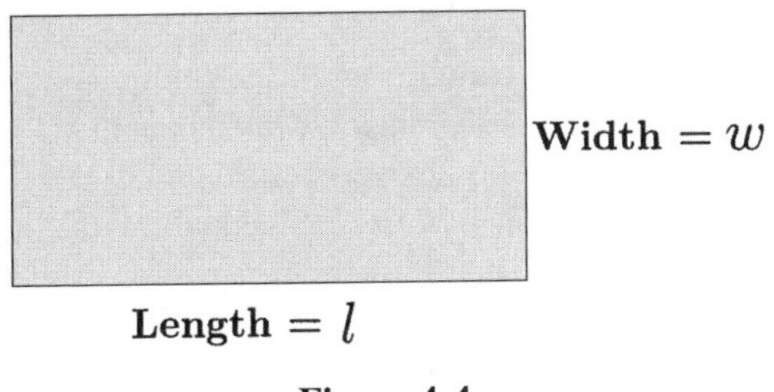

Figure 4-4

Let the length of the rectangle be l and the width be w. The problem tells us that the length is three feet longer than twice the width. Hence, $l = 2w + 3$. Additionally, the perimeter of the rectangle is ninety feet. Recall the formula $P = 2l + 2w$ for the perimeter P of a rectangle with length l and width w. Now write the equation that is applicable for the stated problem and solve it.

$$P = 2l + 2w$$
$$90 = 2(2w + 3) + 2w$$
$$90 = 4w + 6 + 2w$$
$$90 = 6w + 6$$
$$84 = 6w$$
$$14 = w$$

The solution tells us that the width of the rectangle is 14 feet. We were also asked to find the length l. We know that the length = $2w + 3$ so

$$l = 2(14) + 3 = 28 + 3 = 31.$$

Hence, the length of the rectangle is 31 feet.

Solving First-Degree Inequalities

Definition 15. Expressions that utilize the relations $<, \leq, >$, or $\geq$ are called *inequalities.*

Some examples of first-degree inequalities are $3t + 5 > 10$ and $\frac{x-5}{3} \leq 0$.

Definition 16. Any element of the replacement set (domain) of the variable for which the inequality is true is called a *solution.*

Definition 17. The set that contains all of the solutions of an inequality is called the *solution set of the* inequality.

Definition 18. *Equivalent inequalities* are inequalities that have the same solution set.

To solve inequalities, we shall employ a technique that parallels our approach to solving equations. We shall obtain a sequence of equivalent inequalities until we arrive at one whose solution set is obvious. The solution set usually includes one or more intervals when graphed on a number line. Graphing the solution set helps us visualize it.

Properties of Inequalities

If $a < b$, then for real numbers a, b, and q:

(1) $a + q < b + q$ and $a - q < b - q$.

(Adding or subtracting the same quantity on each side of an inequality preserves the order [sense] of the inequality.)

(2) If $q > 0$, $aq < bq$, and $\frac{a}{q} < \frac{b}{q}$.

(Multiplying or dividing both sides of an inequality by a positive quantity preserves the order (sense) of the inequality.)

(3) If $q < 0$, $aq > bq$, and $\frac{a}{q} > \frac{b}{q}$.

(Multiplying or dividing both sides of an inequality by a negative quantity preserves the order (sense) of the inequality.)

Note!

The properties above were stated for the “$<$” relation. The above properties are also valid for the $\leq$, $>$, and $\geq$ relations.

Solved Problem 4.6 Solve the inequality $x - 4 > 2 - x$ and graph the solution set on a number line. Express the solution set using interval notation.

Solution:

$x - 4 > 2 - x$	
$2x > 6$	Add x and 4
$x > 3$	Divide by 2

The solution set in interval notation is $(3, \infty)$. The relevant graph is shown in Figure 4-5.

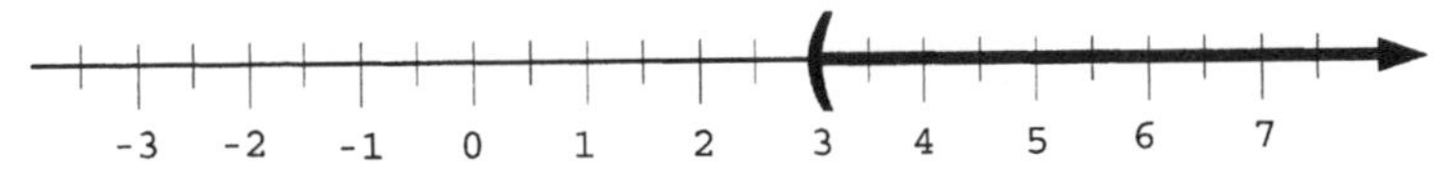

Figure 4-5

Graphs of First-Degree Inequalities

The graphs of first-degree inequalities are a visual display of the solution set of the inequality being discussed. The graph is a region of the coordinate plane called a half-plane. A *half-plane* is the region on either side of a line in a plane.

Solved Problem 4.7 Graph the inequality $x + y \leq 3$.

Solution: Graph the equation $x + y = 3$ using the intercept method. The intercepts occur at (3, 0) and (0, 3). The graph is a line passing through those points. It is the boundary of the half-plane that represents the solution set of the given inequality. Draw the line on the coordinate system. We must now determine the ordered pairs that satisfy the inequality $x + y < 3$. Those ordered pairs are associated with points on one side of the line. We must determine which side is appropriate. Simply identify the coordinates of a point not on the line. Choose the origin if it is not on the boundary. The origin has coordinates (0,0), and $0 + 0 < 3$ is true. We conclude that points on the origin side of the line satisfy the stated inequality. Shade the origin side of the boundary to represent the solution set of the inequality $x + y \leq 3$. See Figure 4-6.

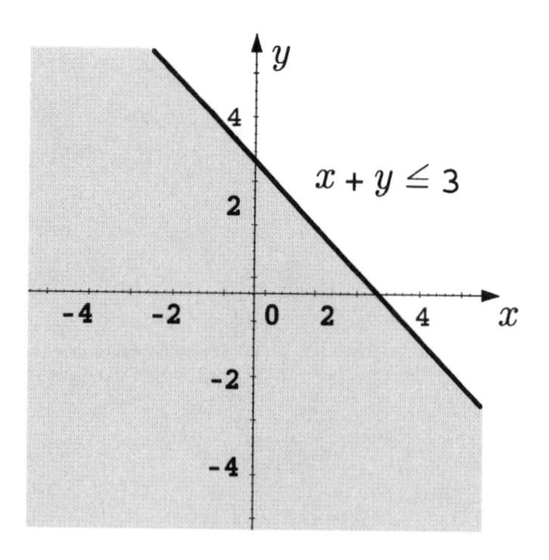

Figure 4-6

The process of graphing linear inequalities is summarized below:

1. Graph the equation that represents the boundary of the solution set. Draw a solid line if the relation involves $\leq$ or $\geq$. Draw a dashed line if the relation involves $<$ or $>$.

2. Choose a convenient test point not on the line and substitute its coordinates into the inequality. Choose a point that has at least one zero coordinate for convenience.

3. If the coordinates of the test point satisfy the inequality, shade the region that contains the point. If the coordinates of the test point do not satisfy the inequality, shade the region on the other side of the line.

Applications Involving Inequalities

Statements that describe inequalities are set up in a manner analogous to those that describe equations. The appropriate relation symbol will be one of the inequality symbols rather than the equals symbol.

Solved Problem 4.8 A retiree requires an annual income of at least \$1500 from an investment that earns interest at 7.5% per year. What is the smallest amount the retiree must invest in order to achieve the desired return?

Solution: Recall that $I = Prt$. Let P be the amount the retiree invests. Then

$$I = Prt \geq 1500$$

$$P\,(0.075)\;1 \geq 1500$$

$$P \geq \frac{1500}{0.075} = 20{,}000$$

The smallest amount the retiree can invest is \$20,000.

Absolute-Value Equations and Inequalities

The absolute value of a real number is the distance between that number and zero on the number line. The equation $|x| = 3$ means that the distance between the points associated with x and 0 on the number line is 3. Hence, x is either -3 or 3 as illustrated in Figure 4-7.

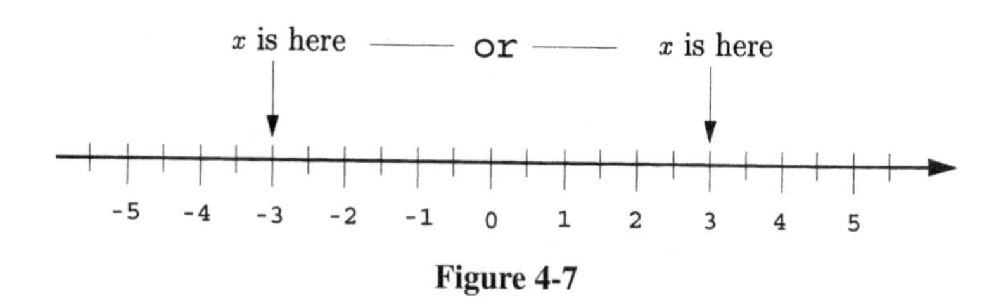

Figure 4-7

Therefore, $x = -3$ and $x = 3$ are solutions to $|x| = 3$. The solution set is $\{-3, 3\}$.

Similarly, $|2x - 1| = 3$ means that the distance between the points associated with $2x - 1$ and 0 is 3. Refer to Figure 4-8.

We conclude that the solutions to $2x - 1 = -3$ and $2x - 1 = 3$ both satisfy $|2x - 1| = 3$. Therefore, $x = -1$ and $x = 2$ are solutions to $|2x - 1| = 3$. The solution set is $\{-1, 2\}$.

Both examples above illustrate that there are two solutions that involve the absolute value of linear expressions that are equal to a positive number. Both solutions are obtained by solving appropriate related equations. We generalize our observations below.

Property 1: If $c \geq 0$, $|q| = c$ is equivalent to $q = -c$ or $q = c$.

Verbally, we say that if c is a nonnegative real number, the solutions to the absolute value of a linear quantity q that is equal to c are the solutions to the equations $q = -c$ and $q = c$. The applications of property 1 permits us to write equations equivalent to the original equation, free of absolute values, which we can solve.

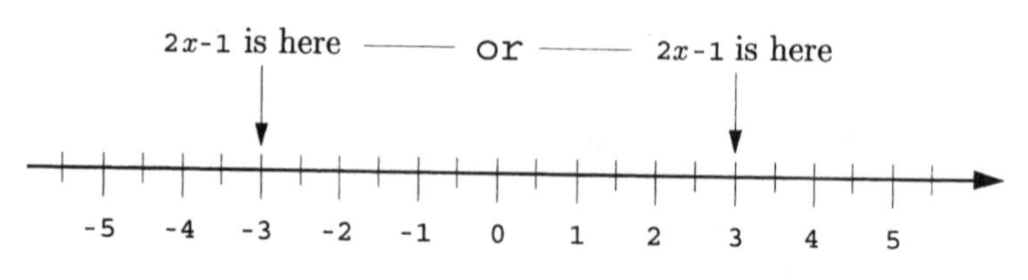

Figure 4-8

Property 2: If $c \geq 0$, $|q| < c$ is equivalent to $-c < q < c$.

Property 3: If $c \geq 0$, $|q| > c$ is equivalent to $q < -c$ or $q > c$.

If the original inequality has the form $|q| \leq c$ or $|q| \geq c$ when $c \geq 0$, property 2 or property 3 is still applicable. We merely replace $<$ by $\leq$ and $>$ by $\geq$ in the associated statements. The solution sets must also include the endpoints of the relevant intervals.

Chapter 5
Exponents, Roots, and Radicals

In This Chapter:

- ✔ *Zero and Negative-Integer Exponents*
- ✔ *Rational Exponents and Roots*
- ✔ *Simplifying Radicals*
- ✔ *Operations on Radical Expressions*

Zero and Negative-Integer Exponents

Zero Exponents

We know that a nonzero quantity divided by itself equals 1. In particular, if n is a positive integer and $b \neq 0$, $b^n/b^n = 1$. Also, if the fourth law of exponents is to hold when $m = n$ and $b \neq 0$, $b^m/b^n = b^n/b^n = b^{n-n} = b^0$. It follows that $b^0 = 1$.

Definition 1. If $b \neq 0$, $b^0 = 1$.

A nonzero quantity raised to the zero power is equal to 1.

The expression 0^0 is indeterminate.

Negative-Integer Exponents

Let us now consider expressions that contain negative-integer exponents. We would like the laws of exponents to hold for negative exponents also. In particular, if $b \neq 0$, $b^n \cdot b^{-n} = b^{n+(-n)} = b^0 = 1$.

The multiplicative inverse property states that if $b \neq 0$, $b^n \cdot 1/b^n = 1$.

Definition 2. If $b \neq 0$, $b^{-n} = 1/b^n$ where n is a positive integer.

In other words, an expression with a negative exponent is equivalent to the reciprocal of that expression with a positive exponent.

Now consider an expression with a negative exponent in the denominator. If n is a natural number (i.e., a positive integer) and $b \neq 0$,

$$\frac{1}{b^{-n}} = \frac{1}{\frac{1}{b^n}}$$

$$= 1 \cdot \frac{b^n}{1}$$

$$= b^n$$

In other words, a nonzero expression with a negative exponent in the denominator is equivalent to the same expression with a positive exponent in the numerator.

Be aware that if $b \neq 0$ and n is a positive integer, the important ideas stated above are:

(i) $b^0 = 1$

(ii) $b^{-n} = \frac{1}{b^n}$

(iii) $\frac{1}{b^{-n}} = b^n$

It can be shown that the laws of exponents stated for positive integers are valid for zero and negative-integer exponents as well. We may apply the laws of exponents in any order. In general, we attempt to choose the order that is most efficient, although the results will be the same regardless of the order chosen. We now present expressions with any integer component in the summary below:

Laws and Definitions for Integer Exponents

Law 1	$b^m \cdot b^n = b^{m+n}$
Law 2	$\left(b^m\right)^n = b^{mn}$
Law 3	$(ab)^n = a^n \cdot b^n$
Law 4	$\frac{b^m}{b^n} = b^{m-n}, b \neq 0$
Law 5	$\left(\frac{a}{b}\right)^n = \frac{a^n}{b^n}, b \neq 0$
Definition 1	$b^0 = 1, b \neq 0$
Definition 2	$b^{-n} = \frac{1}{b^n}, b \neq 0$

The simplest form of an expression with exponents is the form that contains only positive exponents and no base is repeated. When applying Law 4, it is sometimes necessary to apply the definition of negative exponents in order to obtain the result in simplest form.

Rational Exponents and Roots

Our original discussion of exponents required the exponents to be natural numbers. In this section, we give meaning to rational exponents. Recall that a rational number is a number of the form a/b where a and b are integers and $b \neq 0$.

Consider an expression such as $b^{\frac{1}{2}}$. What should $b^{\frac{1}{2}}$ mean? If Law 2 of exponents holds,

$$\left(b^{\frac{1}{2}}\right)^2 = b^{\frac{1}{2} \cdot 2} = b^1 = b$$

Definition 3. $b^{\frac{1}{2}}$ is the nonnegative quantity which, when squared, is equal to b. It is called the *principal square root* of b.

Definition 4. $b^{\frac{1}{3}}$ is the quantity which, when cubed, is equal to b. It is called the *cube root* of b.

Definition 5. $b^{\frac{1}{4}}$ is the nonnegative quantity which, when raised to the fourth power, is equal to b. It is called the *principal fourth root* of b.

Definition 6. If n is a natural number, $b^{\frac{1}{n}}$ is the real number which, when raised to the nth power, is b. It is called the *principal* nth *root* of b.

Definition 7. If $b^{\frac{1}{n}}$ is a real number, then

$$b^{\frac{m}{n}} = \left(b^{\frac{1}{n}}\right)^m = \left(b^m\right)^{\frac{1}{n}}$$

Simplification of expressions that contain rational exponents is analogous to expressions that have integer exponents. The sequence of operations can vary without changing the result. Normally operations within parentheses should be performed first. In other instances, one should do exponentiation first. Analyze the expression to be simplified to determine the most efficient method to employ. In any case, the order of operations stated previously must be adhered to. A particular base should appear as few times as is possible and all exponents should be positive.

We previously stated that $b^{\frac{1}{n}}$ and $\sqrt[n]{b}$ are symbols that represent the nth root of b. $\sqrt[n]{b}$ is the quantity that, when raised to the nth power, yields b. $\sqrt[n]{b}$ is nonnegative if n is even. In $\sqrt[n]{b}$, b is called the *radicand*, $\sqrt{\ }$ is called the *radical sign*, and n is called the *index* of the radical.

Simplifying Radicals

Now that we know how to interpret radicals, we must learn how to simplify them. The definition of a radical and the following three properties provide almost all of the necessary tools.

Properties of Radicals

If $\sqrt[n]{a}$ and $\sqrt[n]{b}$ are real numbers, then:

1. $\sqrt[n]{ab} = \sqrt[n]{a}\sqrt[n]{b}$ (The *n*th root of a product is the product of the *n*th roots of the factors.)

2. $\sqrt[n]{\frac{a}{b}} = \frac{\sqrt[n]{a}}{\sqrt[n]{b}}, b \neq 0$ (The *n*th root of a quotient is the quotient of the *n*th roots of the numerator and denominator provided the denominator is not zero.)

3. $\sqrt[kn]{b^{km}} = \sqrt[n]{b^m}$ for $b \geq 0$ (If the index and the exponent of the radicand contain a common factor, the common factor may be divided out.)

Properties 1 and 2 are actually Laws 3 and 5 of exponents. Property 3 is obtained if we rewrite the radical in exponential form and reduce the exponent. The steps involved are:

$$\sqrt[kn]{b^{km}} = \left(b^{km}\right)^{\frac{1}{kn}} = b^{\frac{km}{kn}} = b^{\frac{m}{n}} = \sqrt[n]{b^m}$$

In order to simplify radical expressions, we must know what is meant by the simplest radical form. The requirements are given below.

Simplest Radical Form

A radical expression is in simplest form if:

1. All factors in the radicand have exponents less than the index.

2. The radicand contains no fractions.

3. No denominator contains a radical.

4. The index and the exponents of the factors in the radicand have no common factor other than one.

You Need to Know 

A perfect *n*th power of a factor must possess an exponent that is a multiple of *n*. Thus, perfect square factors must have exponents that are multiples of two; perfect cube factors must have exponents that are multiples of three; etc. Keep this idea in mind as you simplify radicals.

Solved Problem 5.1 Use the properties of radicals to express the following in simplest radical form: (a) $\sqrt{20}$; (b) $\sqrt[3]{5000}$; and (c) $\sqrt[3]{54x^5}$. For simplicity, assume all variables represent positive numbers.

Solution:

(a) $\sqrt{20} = \sqrt{4 \cdot 5} = \sqrt{2^2 \cdot 5} = \sqrt{2^2} \cdot \sqrt{5} = 2\sqrt{5}$

(b) $\sqrt[3]{5000} = \sqrt[3]{1000 \cdot 5} = \sqrt[3]{10^3 \cdot 5} = \sqrt[3]{10^3}\sqrt[3]{5} = 10\sqrt[3]{5}$

(c) $\sqrt[3]{54x^5} = \sqrt[3]{27 \cdot 2x^3x^2} = \sqrt[3]{3^3x^3}\sqrt[3]{2x^2} = 3x\sqrt[3]{2x^2}$

Rationalizing the Denominator

There are circumstances that result in radicals in the denominator such that none of the factors of the radicand have exponents that are multiples of the index. This situation violates criterion number 3 for simplifying radicals. We apply the fundamental principle of fractions to obtain a rational expression in the denominator. This process is called *rationalizing the denominator*. The technique employed is to multiply the numerator and denominator by the expression that changes the denominator into the *n*th root of a perfect *n*th power. That is, the denominator is a rational expression if it is the square

root of a perfect square, or the cube root of a perfect cube, or, in general, the nth root of a perfect nth power.

Solved Problem 5.2 Assuming all radicands are positive numbers, implify the following: (a) $\frac{4}{\sqrt{5}}$; (b) $\frac{7}{3\sqrt{7}}$; and (c) $\frac{8\sqrt{2}}{\sqrt{5}}$.

Solution:

(a) $\frac{4}{\sqrt{5}}$

We must obtain a perfect square in the radicand since the index is 2. Multiply the numerator and denominator by $\sqrt{5}$ to accomplish the task.

$$\frac{4}{\sqrt{5}} = \frac{4}{\sqrt{5}} \cdot \frac{\sqrt{5}}{\sqrt{5}} = \frac{4\sqrt{5}}{\sqrt{5^2}} = \frac{4\sqrt{5}}{5}$$

(b) $\frac{7}{3\sqrt{7}}$

We must obtain a perfect square in the radicand since the index is 2. Multiply the numerator and denominator by $\sqrt{7}$ to accomplish the task.

$$\frac{7}{3\sqrt{7}} = \frac{7}{3\sqrt{7}} \cdot \frac{\sqrt{7}}{\sqrt{7}} = \frac{7\sqrt{7}}{3\sqrt{7^2}} = \frac{7\sqrt{7}}{3 \cdot 7} = \frac{\sqrt{7}}{3}$$

(c) $\frac{8\sqrt{2}}{\sqrt{5}}$

We must obtain a perfect square in the radicand since the index is 2. Multiply the numerator and denominator by $\sqrt{5}$ to accomplish the task.

$$\frac{8\sqrt{2}}{\sqrt{5}} = \frac{8\sqrt{2}}{\sqrt{5}} \cdot \frac{\sqrt{5}}{\sqrt{5}} = \frac{8\sqrt{2 \cdot 5}}{\sqrt{5^2}} = \frac{8\sqrt{10}}{5}$$

Operations on Radical Expressions

Adding and Subtracting

The distributive properties, $a\,(b + c) = ab + ac$ and $(b + c)a = ba + ca$, were used previously to combine like terms. These same properties can be employed to combine like radicals. *Like radicals* are radicals with the same indices and radicands. Like radicals can be manipulated just as like terms can be manipulated.

Solved Problem 5.3 Assuming variables represent positive numbers, simplify $3\sqrt{5} + 8\sqrt{5}$.

Solution:

$$3\sqrt{5} + 8\sqrt{5} = (3 + 8)\sqrt{5} = 11\sqrt{5}$$

Multiplying and Dividing

We employ the distributive properties to multiply expressions that contain radicals. The process is analogous to multiplying polynomials. Radicals multiplied by factors should be written with the factors preceding the radical in order to avoid ambiguity. That is, write $y\sqrt{x}$ rather than $x\sqrt{y}$.

Solved Problem 5.4 Assuming radicands are positive, simplify $\sqrt{3}\left(\sqrt{2} - \sqrt{3}\right)$.

Solution:

$$\sqrt{3}\left(\sqrt{2} - \sqrt{3}\right) = \sqrt{3}\sqrt{2} - \sqrt{3}\sqrt{3} = \sqrt{6} - \sqrt{9} = \sqrt{6} - 3$$

Chapter 6
Second-Degree Equations and Inequalities

In This Chapter:

- ✔ *Solving by Factoring and Square Root Methods*
- ✔ *Completing the Square and the Quadratic Formula*
- ✔ *Equations Involving Radicals*
- ✔ *Quadratic Form Equations*
- ✔ *Graphs of Second-Degree Equations*
- ✔ *Quadratic and Rational Inequalities*

Solving by Factoring and Square Root Methods

A *second-degree equation* is a polynomial equation of degree two. Second-degree equations in one variable are commonly referred to as *quadratic* equations. The *standard form* of a quadratic equation is $ax^2 + bx + c = 0$ where $a > 0$. The *solutions*, *roots*, or *zeros* of a quadratic equation

are the values of the variable for which the equation is true. There are normally two solutions or two roots of a quadratic equation, although there occasionally is one or no solution to the equation.

The variable terms may vanish as we solve certain equations. The equation $x^2 + 1 = x^2 - 2$ has no solution since the x^2 terms vanish if we subtract x^2 from both sides and obtain $1 = -2$. This statement is a contradiction. There is no solution to the equation.

If a given equation contains a variable in the denominator of a fraction, we multiply by the LCD to clear all fractions. The resulting equation may have solutions that do not satisfy the original equation. This situation occurs for values of the variable that produce a zero value in the original denominator. Equivalently, the LCD may contain a factor whose value is zero for a particular value of the variable. Recall that equivalent equations are obtained only when we multiply both sides of an equation by a nonzero quantity. The apparent solutions to the resulting equation that do not satisfy the original equation are called *extraneous roots*. We are obligated to check for extraneous roots when we multiply factors that involve a variable. It is part of the solution process.

Factor Method

The reader should review factoring polynomials in Chapter 2 before continuing. We shall employ the zero factor property to solve quadratic equations.

Zero Factor Property: If $a \cdot b = 0$, then $a = 0$ or $b = 0$.

The zero factor property states that if the product of two quantities is zero, then at least one of the quantities is zero.

Solved Problem 6.1 Using the factor method, solve the quadratic equation

$$x^2 + x - 12 = 0$$

Solution: Factor the quadratic expression and apply the zero factor property:

$$x^2 + x - 12 = (x + 4)(x - 3) = 0$$

Now solve the resulting linear equations.

$$x + 4 = 0 \qquad\qquad x - 3 = 0$$
$$x = -4 \qquad\qquad x = 3$$

The solution set is $\{-4, 3\}$.

Square Root Method

If $b = 0$ in the quadratic equation $ax^2 + bx + c = 0$, the equation becomes $ax^2 + c = 0$. Furthermore, if c is subtracted from both sides and both sides are divided by a, then $x^2 = -c/a$. Replace $-c/a$ by the single letter d. The equation then has the form $x^2 = d$. The values of x for which the equation is true are those numbers we can square to obtain d. The possibilities are $\sqrt{d}$ and $-\sqrt{d}$. A common way to express $\sqrt{d}$ and $-\sqrt{d}$ is $\pm\sqrt{d}$. The results of the above discussion are stated in the theorem that follows.

Theorem 1: If $x^2 = d$, then $x = \pm\sqrt{d}$. The solution set is $\{\pm\sqrt{d}\}$ or $\{-\sqrt{d}, +\sqrt{d}\}$.

If the first-degree term is missing in a quadratic equation, we may find the roots by applying the above theorem. This technique is referred to as the square root method or simply as the extraction of roots. The square root method may also be employed to solve equations that involve a perfect square equaling a real number.

Solved Problem 6.2 Using the square root method, solve the quadratic equation

$$x^2 - 49 = 0$$

Solution: Solve for x^2 and extract roots.

$$x^2 - 49 = 0$$
$$x^2 = 49$$
$$x = \pm\sqrt{49} = \pm 7$$

The solution set is $\{\pm 7\} = \{-7, +7\}$.

Completing the Square and the Quadratic Formula

Quadratic equations are best solved by factoring but not all quadratic expressions factor. Consider $x^2 + x + 1$, for example. Additionally, the square root method only works if the equation can be written in the form $(x+k)^2 = d$.

We now develop a technique that works for *all* quadratic equations. The process is rather cumbersome, although it is important in a variety of applications you may encounter in subsequent courses. The process is called *completing the square*.

We first observe some useful relationships when a binomial of the form $x + k$ is squared. The result is a perfect square trinomial:

$$(x+k)^2 = x^2 + 2kx + k^2$$

You Need to Know 

Note the following relationships on the right side of the equation.

1. The coefficient of the squared term, x^2, is one.
2. The coefficient of the linear term, x, is $2k$.
3. The constant term, k^2, is the square of one-half the coefficient of x.

Completing the Square

Our equation must be in the form $(x+k)^2 = d$ in order for us to solve by the square root method. Hence our first objective will be to use the above skills to write the equation in the appropriate form. Our next step is to extract roots.

The steps involved in solving a quadratic equation by completing the square are summarized below:

1. Isolate the variable terms.
2. Make the coefficient of the squared term 1 if it is other than 1.
3. Determine the square of one-half the coefficient of the linear term and add to both sides.
4. Factor the perfect square trinomial.
5. Extract square roots and simplify if necessary.
6. Isolate the variable.
7. State the solution set.

The Quadratic Formula

The quadratic formula is developed through solving a quadratic equation by completing the square. We begin with the standard form of the equation

$$ax^2 + bx + c = 0$$

1. Isolate the variable terms. $\quad ax^2 + bx = -c$

2. Divide both sides by a. $\quad x^2 + \frac{b}{a}x = \frac{-c}{a}$

3. Determine $k^2 = \left[\frac{1}{2}\left(\frac{b}{a}\right)\right]^2 = \frac{b^2}{4a^2}$

Add to both sides and simplify.

$$\begin{aligned} x^2 + \frac{b}{a}x + \frac{b^2}{4a^2} &= \frac{-c}{a} + \frac{b^2}{4a^2} \\ &= \frac{-c}{a} \cdot \frac{4a}{4a} + \frac{b^2}{4a^2} \\ &= \frac{b^2 - 4ac}{4a^2} \end{aligned}$$

4. Factor the left side. $\left(x+\frac{b}{2a}\right)^2=\frac{b^2-4ac}{4a^2}$

5. Extract square roots and simplify.

$$x+\frac{b}{2a}=\pm\sqrt{\frac{b^2-4ac}{4a^2}}=\pm\frac{\sqrt{b^2-4ac}}{\sqrt{4a^2}}=\pm\frac{\sqrt{b^2-4ac}}{2a}$$

6. Isolate the variable.

$$x=-\frac{b}{2a}\pm\frac{\sqrt{b^2-4ac}}{2a}=\frac{-b\pm\sqrt{b^2-4ac}}{2a}$$

The last equation is called the quadratic formula.

You Need to Know 

The Quadratic Formula

The solutions to $ax^2+bx+c=0$, $a\neq 0$, are given by

$$x=\frac{-b\pm\sqrt{b^2-4ac}}{2a}$$

Solved Problem 6.3 Using the quadratic formula, solve the equation

$$x^2-3x+1=0$$

Solution: The equation is in the appropriate form. Therefore, the coefficients can readily be identified: $a=1$, $b=-3$, and $c=1$. Substitute the values into the formula and simplify.

$$x=\frac{-b\pm\sqrt{b^2-4ac}}{2a}=\frac{-(-3)\pm\sqrt{(-3)^2-4(1)(1)}}{2(1)}=\frac{3\pm\sqrt{9-4}}{2}=\frac{3\pm\sqrt{5}}{2}$$

The solution set is $\left\{\frac{3 \pm \sqrt{5}}{2}\right\}$.

In summary, to solve quadratic equations:

1. Use the factor method if the expression can be factored rather easily.
2. If the first-degree term is missing, employ the square root method.
3. The quadratic formula may be utilized to solve other quadratic equations. It may be employed to solve all quadratic equations, although it is less efficient for the types suggested in steps 1 and 2 above.
4. Completing the square may be used without exception, although it is the most cumbersome and prone to error. Completing the square is an important technique that we need for reasons other than solving equations.

Equations Involving Radicals

If a given equation contains one or more radical expressions, we shall attempt to eliminate the radicals to solve the equation. The property that facilitates the process is stated below.

Property 1. If $a = b$, then $a^2 = b^2$.

Essentially, the property says if two expressions are equal, then their squares are equal. The statement cannot be reversed, however. That is, the converse is not true. We'll make use of the fact that the solutions of the equation $a = b$ are included in the solutions of $a^2 = b^2$. The squared version sometimes has solutions that are not solutions of the original equation. In other words, squaring both sides of an equation sometimes introduces extraneous roots. We are obligated to check for extraneous roots when property 1 is used; it is part of the solution process.

Property 1 can be generalized as follows:

Property 2. If $a = b$, then $a^p = b^p$ where p is a rational number.

In other words, rational powers of equal expressions are equal. If the rational exponent is in reduced form, and its denominator is even, a check is required as part of the solution process. Property 2 is used to solve equations that contain higher order roots.

Quadratic Form Equations

Some equations can be transformed into a quadratic equation by substitution. We shall let the variable u represent an appropriate expression, then substitute and solve an equation of the form $au^2 + bu + c = 0$. Once values for u are obtained, we will find values for the original variable that satisfy the given equation. Quadratic form equations are encountered in equations in which the variable factor in one term is the square of the variable factor in another term and no other terms contain variable factors.

Solved Problem 6.4 Solve the equation

$$y^4 - 9y^2 + 20 = 0$$

Solution: Since y^4 is the square of y^2, let $u = y^2$ so that $u^2 = y^4$. Now substitute to obtain

$$u^2 - 9u + 20 = 0$$
$$(u-5)(u-4) = 0$$

$$u = 5 \quad \text{or} \quad u = 4$$

Since $u = y^2$,

$$y^2 = 5 \quad \text{or} \quad y^2 = 4$$
$$y = \pm\sqrt{5} \quad \text{or} \quad y = \pm 2$$

The solution set is $\{\pm\sqrt{5}, \pm 2\}$.

Graphs of Second-Degree Equations

The graph of $ax^2 + bx + c = 0,\ a \neq 0$, is a curve called a *parabola*. Since the curve is not a straight line, we must plot more points in order to establish the pattern that is necessary when graphing first-degree equations.

The shape of the parabola will become apparent as we graph the following equations.

Solved Problem 6.5 Graph the following equations: (*a*) $y = x^2$ and (*b*) $y = x^2 + 2$.

Solution: We begin by choosing arbitrary values for x and determine the corresponding y values. Make a table of values. Plot the points and connect them with a smooth curve.

(*a*)

x	y
0	0
1	1
2	4
3	9
−1	1
−2	4
−3	9

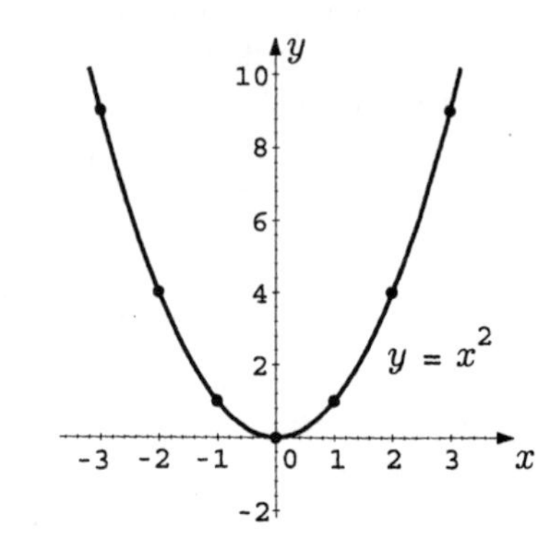

(*b*)

x	y
0	2
1	3
2	6
3	11
−1	3
−2	6
−3	11

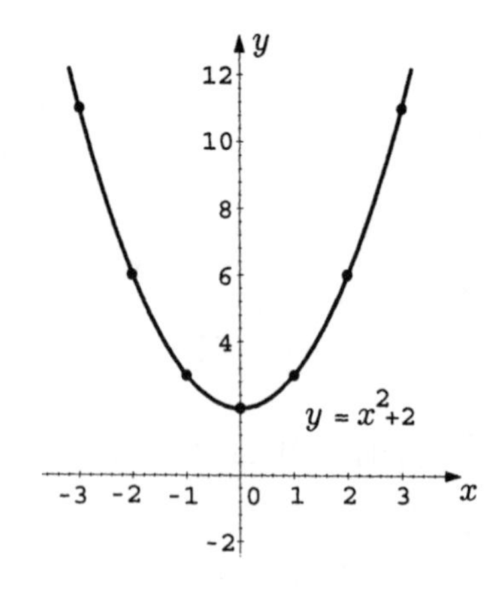

Observe that the parabola opens upward when a, the coefficient of x^2, is positive. The parabola opens downward when a is negative. We can therefore predict the direction the parabola opens merely by observing the signs of the coefficient of the x^2 term.

We now discuss how we can determine the more significant parts of the graph. The highest or lowest point on the curve can be very important. The highest or lowest point on the curve is called the *maximum* or *minimum point*, respectively.

The maximum or minimum point on the graph of $y = ax^2 + bx + c$, $a \neq 0$, occurs at $x = -b/(2a)$. The maximum or minimum point is called the *vertex* of the parabola. Evaluate $y = ax^2 + bx + c$ at $x = -b/(2a)$ to find the y-coordinate of the vertex.

A line parallel to the y-axis that passes through the vertex divides the parabola into parts that are mirror images of each other. This line is called the *axis of symmetry* of the parabola. The idea of symmetry can be employed when graphing the curve. Recall that the y-intercept of a graph is found if we let $x = 0$ and solve for y. Similarly, if $y = 0$, the value(s) of x are the x-intercepts. In part (a) above, the y-intercept is 5 and the x-intercepts are 1 and 5. The intercepts of a parabola should usually be found. Sometimes the x-intercepts are irrational and therefore cumbersome to determine. We normally omit x-intercepts if this is the case. The ideas discussed above are summarized below.

To graph parabolas with the equation $y = ax^2 + bx + c$:

1. If $a > 0$, the parabola opens upward. If $a < 0$, the parabola opens downward.
2. Find the x-coordinate of the vertex. It is given by $x = -b/(2a)$.
3. Find the y-coordinate of the vertex. Evaluate $y = ax^2 + bx + c$ at $x = -b/(2a)$.
4. Let $x = 0$ and evaluate $y = ax^2 + bx + c$ to determine the y-intercept.

5. Let $y = 0$ and solve $y = ax^2 + bx + c$ for x to determine the x-intercepts, if any. This step is optional if the expression does not factor.

6. Plot the points found and one or two additional points. Use symmetry to complete the graph.

Solved Problem 6.6 Utilizing the procedure stated above, graph the equation $y = x^2 - 4x$.

Solution:

1. The parabola opens upward since $a = 1 > 0$.
2. The vertex occurs at $x = \dfrac{-b}{2a} = \dfrac{-(-4)}{2(1)} = \dfrac{4}{2} = 2$.
3. The y-coordinate of the vertex is $y = 2^2 - 4(2) = 4 - 8 = -4$.
4. If $x = 0$, $y = 0^2 - 4(0) = 0 - 0 = 0$. This is the y-intercept.
5. If $y = 0$, $0 = x^2 - 4x$ or $x^2 - 4x = 0$. Hence, $x(x - 4) = 0$.

Therefore, $x = 0$ and $x = 4$ are the x-intercepts. We record the ordered pairs found in a table and find a couple of additional points to graph. Plot the points found and draw a smooth curve through them to complete the graph.

x	y
2	−4
0	0
4	0
−1	5
5	5

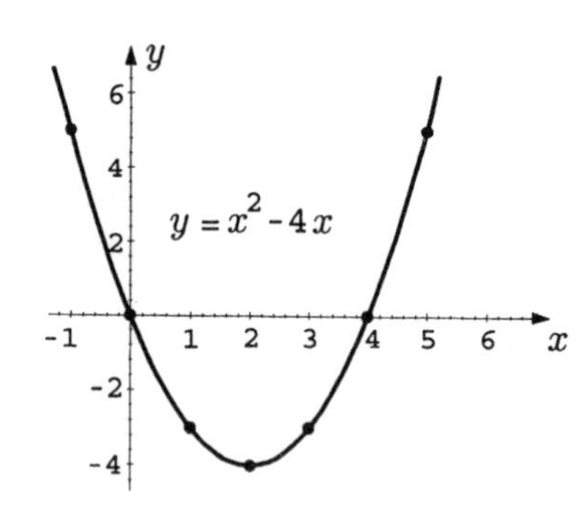

Quadratic and Rational Inequalities

Quadratic Inequalities

A quadratic inequality of the form, $ax^2 + bx + c < 0$, $a > 0$, is in *standard form*. (Note that $<$ may be replaced by $>$, $\leq$, or $\geq$.)

In order to solve the inequality $x^2 - x - 6 < 0$, we must find the values of x for which $x^2 - x - 6$ is negative. We begin by solving the equation $x^2 - x - 6 = 0$. The factor method works well.

$$x^2 - x - 6 = (x+2)(x-3) = 0 \quad \text{if } x = -2 \text{ or } x = 3.$$

The value(s) of the variable for which an expression is either zero or is undefined are called the *critical values(s)* for the expression. Thus −2 and 3 are the critical values for the expression $x^2 - x - 6$.

The inequality $x^2 - x - 6 < 0$ can be written as $(x+2)(x-3) < 0$. The product of the factors is negative when one factor is positive and the other factor is negative. We shall examine the signs of the factors $x + 2$ and $x - 3$ for various arbitrary values in the intervals established by the critical values −2 and 3.

We illustrate the relevant information in Figure 6-1.

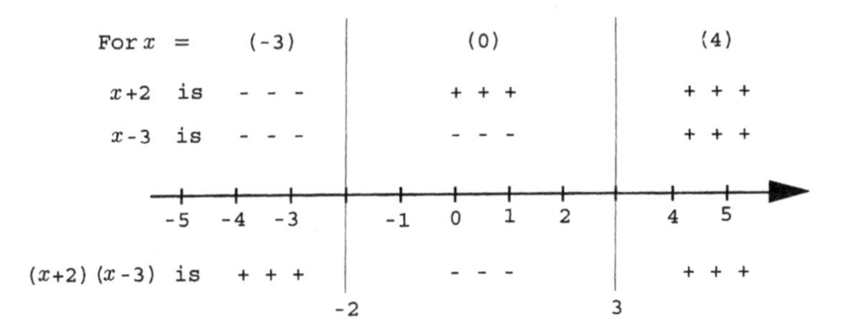

Figure 6-1

We call the figure shown in Figure 6-1 a *sign diagram.* The signs of the factors and their product for various values of x in the intervals formed by graphing the critical values on a number line are shown. The value for x (test value) used in a given interval is arbitrary. (Choose an integer value to test in general.) We also note that a factor has the same sign for all values of the variable in a particular interval. We can determine the solution set by reading the results indicated from the sign diagram.

In our illustration, it is apparent that $x^2 - x - 6 < 0$ for all x in the interval (−2, 3). We now summarize the procedure employed above.

Procedure for Solving Quadratic Inequalities

1. Write the inequality in standard form.
2. Find the critical values by forming a quadratic equation of the form $ax^2 + bx + c = 0$. Solve by factoring or use the quadratic formula.
3. Construct a sign diagram for the factors by choosing arbitrary test values for the variable in the intervals determined by the critical values. Find the signs of the factors and their product in each interval.
4. Identify and record the solution set for the inequality.

Solved Problem 6.7 Solve the inequality $x^2 + 3x - 10 > 0$, making a sign diagram and expressing the solution set in interval notation.

Solution:

We must first find the values of x for which the expression is positive. Employ the procedure stated above.

1. $x^2 + 3x - 10 > 0$.

2. Solve $x^2 + 3x - 10 = 0$. $x^2 + 3x - 10 = (x+5)(x-2) = 0$ if $x = -5$ or $x = 2$. The critical values are -5 and 2.

3.

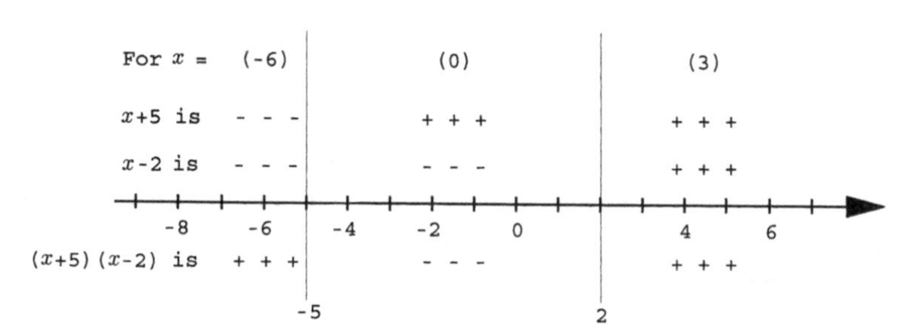

4. The solution set is $(-\infty, -5) \cup (2, \infty)$.

Chapter 7
SYSTEMS OF EQUATIONS AND INEQUALITIES

IN THIS CHAPTER:

- ✔ *Linear Systems in Two Variables*
- ✔ *Linear Systems in Three Variables*
- ✔ *Determinants and Cramer's Rule*
- ✔ *Matrix Methods*
- ✔ *Nonlinear System*
- ✔ *Systems of Inequalities*

Linear Systems in Two Variables

Equations of the form $ax + by = c$, where a and b are not both zero, and $dx + ey = f$, where d and e are not both zero, are linear equations. If two or more linear equations are considered together, a *linear system* is formed. We write

$$\begin{cases} ax + by = c \\ dx + ey = f \end{cases}$$

to represent the system.

Sometimes one of the equations may have the form $y = ax + b$. In that case, the system may have the form

$$\begin{cases} y = ax + b \\ cx + dy = f \end{cases}$$

The brace tells us to consider the equations together. The system is a 2×2 (read 2 by 2), system since there are two equations and two variables.

Solving a system of equations consists of finding all the ordered pairs, if any, which satisfy both equations in the system simultaneously.

How many ordered pairs can satisfy a system of two equations and two unknowns? The answer is clear if we look at graphs that represent the several possibilities.

Case 1:

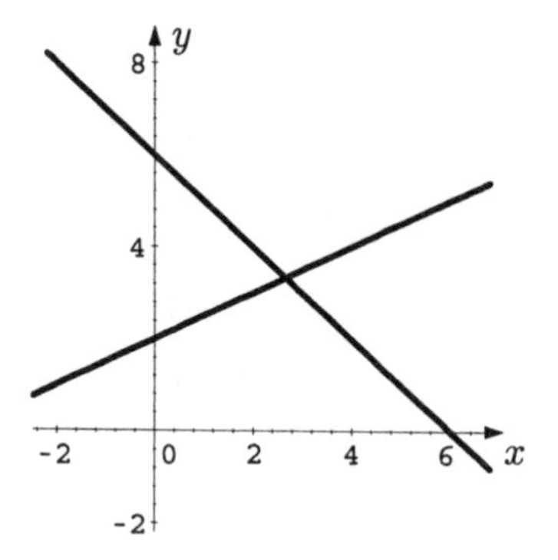

Figure 7-1

The lines intersect in exactly one point. The ordered pair that specifies the coordinates of the point of intersection of the lines is the solution to the system. The solution set contains that ordered pair as its only element. This system is said to be *consistent* and *independent*.

Case 2:

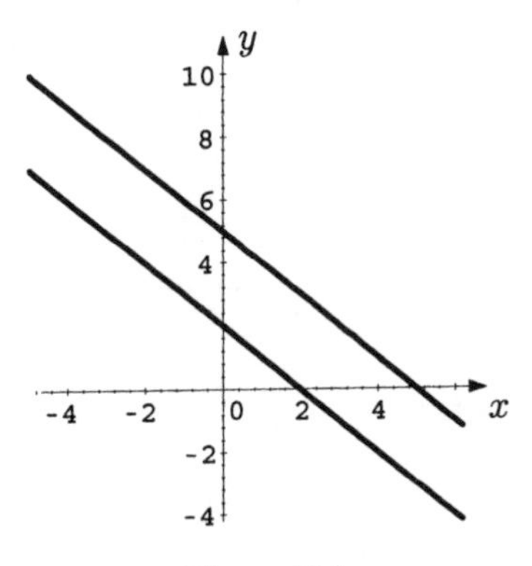

Figure 7-2

The lines are parallel; they do not intersect. There is no solution to the system. The solution set is the empty set ∅. This system is said to be *inconsistent*.

Case 3:

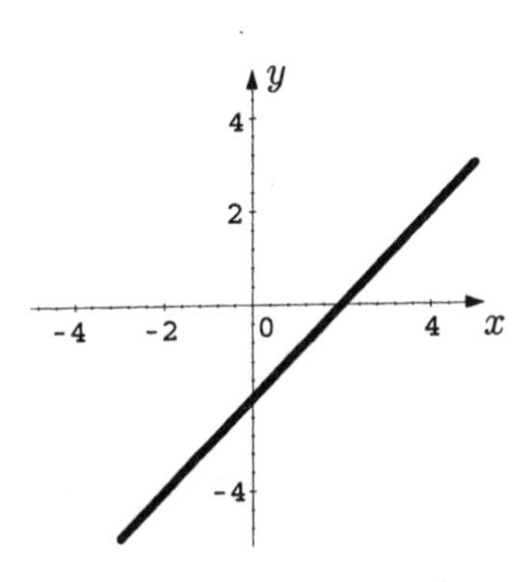

Figure 7-3

The equations represent the same line. All the points on one line are also on the other. There are infinitely many solutions to the system. The solution set contains infinitely many ordered pairs. This system is said to be *dependent.*

We now discuss algebraic methods that allow us to determine the solution or solutions, if any, to a linear system. The methods employed essentially eliminate unknowns until one equation in one unknown is obtained that we can solve. The result is then used to find the remaining unknown.

The Addition Method

The addition method involves the manipulation of the equations to obtain one equation in one unknown when equations are added. We ordinarily multiply one or both equations by suitable factors to accomplish the task. The appropriate factors are those that result in additive inverse coefficients of one variable in both equations. The process is illustrated below.

Solved Problem 7.1 Use the addition method to solve the following system of equations:

$$\begin{cases} 2x - y = 4 \\ \quad\quad\; y = 6 \end{cases}$$

Solution: The second equation states the value of y directly. We need only to substitute $y = 6$ into the first equation and solve for x.

$$2x - y = 4$$
$$2x - 6 = 4$$
$$2x = 10 \Rightarrow x = 5$$

The solution is (5, 6). The solution set is $\{5, 6\}$. The equations are consistent and independent.

The addition method of solving linear systems in two variables is summarized below:

1. Write the equations in the standard form $ax + by = c$.
2. Determine which variable you wish to eliminate.

3. Multiply one or both equations by the appropriate constant or constants to obtain coefficients of the chosen variable that are additive inverses.

4. Add the equations obtained in step 3.

5. Solve the equation in one variable obtained in step 4.

6. Substitute the value of the variable obtained in step 5 in either of the original equations and solve for the other variable.

7. Express the solution as an ordered pair or as a solution set.

8. Check your solution in the original equations.

The Substitution Method

The substitution method entails solving one equation for one variable in terms of the other. We can avoid fractions if we solve for a variable that has a coefficient of +1 or −1. We then substitute the expression obtained into the remaining equation. The process results in one equation with one unknown that we can solve. We then proceed in a manner analogous to the addition method.

Solved Problem 7.2 Use the substitution method to solve the following system of equations:

$$\begin{cases} x - 3y = -7 \\ 4x + 3y = 2 \end{cases}$$

Solution: Solve $x - 3y = -7$ for x in terms of y:

$$x - 3y = -7 \Rightarrow x = 3y - 7$$

Now substitute $3y - 7$ for x into $4x + 3y = 2$ and solve for y.

$$4(3y - 7) + 3y = 2$$
$$12y - 28 + 3y = 2$$
$$15y = 30 \Rightarrow y = 2$$

Now substitute 2 for y into $x = 3y - 7$ and solve for x:

$$x = 3(2) - 7 = 6 - 7 = -1$$

The solution is $(-1, 2)$. The solution set is $\{-1, 2\}$. The system is consistent and independent.

The substitution method of solving linear systems in two variables is summarized below:

1. Solve one of the equations for one variable in terms of the other. Clear fractions first if necessary.
2. Substitute the expression obtained in step 1 into the other equation.
3. Solve the equation obtained in step 2 for the unknown.
4. Substitute the numerical value obtained in step 3 for one variable into either equation in two variables and solve for the remaining variable.
5. State the solution or solution set.
6. Check the solution in each of the original equations.

Although we can use either of the methods discussed to solve linear systems of two equations in two unknowns, circumstances dictate which method may be the most efficient. As a general rule, if no variable has a coefficient of +1 or −1 in either equation, the addition method is preferred. Both techniques have advantages and disadvantages.

Linear Systems in Three Variables

A solution of an equation in three variables, such as $x - 2y + 3z = -4$, is an ordered triplet of real numbers (x, y, z). Values for each of the three variables must be substituted into the equation before we can decide whether the result is a true statement. The values of the components in the ordered triplet are listed in alphabetical order. Thus, $(3, 2, -1)$ and $(-4, 0, 0)$ are solutions of the equation, while $(1, 0, 1)$ is not. There are infinitely many ordered triplets in this solution set.

The solution set of a system of three linear equations in three unknowns, such as

$$\begin{cases} x - 2y + 3z = -4 \\ 2x - y - z = 5 \\ 3x + 2y + z = 12 \end{cases}$$

is the intersection of the solution sets of the individual equations in the system.

The system is a 3×3 (read 3 by 3) system since it consists of three equations in three unknowns. The graph of a first-degree equation in two variables is a straight line in a two-dimensional coordinate system. The graph of a first-degree equation in three variables is a plane in a three-dimensional coordinate system. Each equation in a 3×3 system therefore represents a plane in three-dimensional space. Three planes may intersect at a unique point, may intersect at infinitely many points, or may not intersect at points common to all three planes. There are therefore three possibilities for the solution set of a 3×3 system:

1. There is a unique solution of the system.
2. There are infinitely many solutions of the system.
3. There is no solution of the system.

We employ methods similar to those used to solve systems of linear equations in two variables to solve a 3×3 system. We eliminate variables from pairs of equations until an equation in one unknown is obtained. We then employ "back substitution" to find remaining unknowns. We illustrate by solving the system given above:

$$\begin{cases} x - 2y + 3z = -4 & (1) \\ 2x - y - z = 5 & (2) \\ 3x + 2y + z = 12 & (3) \end{cases}$$

Look at the system and choose a variable to eliminate first. Sometimes one variable can be eliminated more easily than the others. We shall eliminate y first using equations (1) and (3).

$$\begin{array}{rl} x - 2y + 3z = -4 & \\ 3x + 2y + z = 12 & \text{Add} \\ \hline 4x + 0 + 4z = 8 & (4) \end{array}$$

Now choose two different equations and eliminate y again. We shall use (2) and (3).

$$\begin{array}{rrl} \text{2 times (2)} \rightarrow & 4x - 2y + 2z = 10 & \\ & 3x + 2y + z = 12 & \text{Add} \\ \hline & 7x + 0 - z = 22 & (5) \end{array}$$

We now form a 2×2 system that consists of equations (4) and (5). Use the methods of the previous section to solve the system.

$$\begin{cases} 4x + 4z = 8 & (4) \\ 7x - z = 22 & (5) \end{cases}$$

$$\begin{array}{rll} \frac{1}{4} \text{ times } (4) \rightarrow & x + z = 2 & \\ & 7x - z = 22 & \text{Add} \\ \hline & 8x + 0 = 24 \Rightarrow x = 3 & \end{array}$$

"Back substitute" by substituting $x = 3$ into (4) or (5) to find $z = -1$. Now substitute $x = 3$ and $z = -1$ into one of the original equations to find y. We choose equation (2) and find $y = 2$. The solution is $x = 3$, $y = 2$, and $z = -1$ or $(3, 2, -1)$. The solution set is $\{3, 2, -1\}$.

The procedure is summarized below. To solve a linear system in three variables:

1. Eliminate any variable from any pair of equations. Try to choose the most convenient variable to eliminate if there is one.
2. Use appropriate steps to eliminate the *same variable* as in step 1 from a *different* pair of the original equations.
3. Solve the resulting system of two equations in two variables.
4. Substitute the values obtained in step 3 into one of the original equations. Solve for the remaining variable.
5. Check the solution in *each* of the original equations.

If the resulting equation vanishes or yields a contradiction at any step in the process, the system contains dependent equations or else two or three inconsistent equations. The system then has infinitely many solutions or no solution, respectively.

Determinants and Cramer's Rule

Determinants

The process of solving systems of linear equations is rather cumbersome. Fortunately, the procedure is repetitive in nature. We shall introduce a technique that facilitates the process through the use of determinants. A *determinant* is a number that is associated with a square array of numbers.

Definition 1. A 2×2 *determinant*, designated by $\begin{vmatrix} a & b \\ c & d \end{vmatrix}$, has the value $ad - bc$. It is called a *second-order determinant.*

Solved Problem 7.3 Evaluate the following determinants:

(a) $\begin{vmatrix} 2 & 3 \\ 1 & 5 \end{vmatrix}$, (b) $\begin{vmatrix} -3 & -2 \\ 4 & 5 \end{vmatrix}$, and (c) $\begin{vmatrix} 5 & 6 \\ 0 & -2 \end{vmatrix}$.

Solution:

(a) $\begin{vmatrix} 2 & 3 \\ 1 & 5 \end{vmatrix} = 2(5) - 3(1) = 10 - 3 = 7$

(b) $\begin{vmatrix} -3 & -2 \\ 4 & 5 \end{vmatrix} = -3(5) - (-2)(4) = -15 - (-8) = -15 + 8 = -7$

(c) $\begin{vmatrix} 5 & 6 \\ 0 & -2 \end{vmatrix} = 5(-2) - 6(0) = -10 - 0 = -10.$

A third-order determinant has three rows and three columns. Its value is defined as follows:

Definition 2. A 3×3 *determinant*, designated by $\begin{vmatrix} a_1 & b_1 & c_1 \\ a_2 & b_2 & c_2 \\ a_3 & b_3 & c_3 \end{vmatrix}$, has value

$$a_1b_2c_3 + b_1c_2a_3 + c_1a_2b_3 - a_3b_2c_1 - b_3c_2a_1 - c_3a_2b_1.$$

Because this definition is very cumbersome and difficult to remember, we evaluate third-order determinants by a method called *expansion by minors*. The *minor of an element* (number or letter) of a 3×3 determinant is the 2×2 determinant that remains after you delete the row and column which contain the element. Therefore, to evaluate a 3×3 determinant by expansion by minors of elements in the first column, apply the following relationship:

$$\begin{vmatrix} a_1 & b_1 & c_1 \\ a_2 & b_2 & c_2 \\ a_3 & b_3 & c_3 \end{vmatrix} = a_1\begin{vmatrix} b_2 & c_2 \\ b_3 & c_3 \end{vmatrix} - a_2\begin{vmatrix} b_1 & c_1 \\ b_3 & c_3 \end{vmatrix} + a_3\begin{vmatrix} b_1 & c_1 \\ b_2 & c_2 \end{vmatrix}$$

Evaluate the 2×2 determinants as illustrated previously. Notice that the signs of the coefficients of the terms alternate. The result is unchanged if we expand the determinant by minors of elements in any column or row. The appropriate signs of the terms in the expansion are displayed by position in the array below:

$$\begin{matrix} + & - & + \\ - & + & - \\ + & - & + \end{matrix}$$

Cramer's Rule

Cramer's rule may be employed to solve systems of linear equations. His rule expresses the solution for each variable as the quotient of two determinants. This allows us to readily employ computers to solve systems of linear equations.

Cramer's Rule for 2 × 2 Systems

The solution to $\begin{cases} a_1x + b_1y = k_1 \\ a_2x + b_2y = k_2 \end{cases}$ is given by $x = D_x/D$ and $y = D_y/D$

where $D = \begin{vmatrix} a_1 & b_1 \\ a_2 & b_2 \end{vmatrix}$ and $D \neq 0$, $D_x = \begin{vmatrix} k_1 & b_1 \\ k_2 & b_2 \end{vmatrix}$ and $D_y = \begin{vmatrix} a_1 & k_1 \\ a_2 & k_2 \end{vmatrix}$.

In practice, find D first since there is no unique solution if $D = 0$.

Cramer's rule can be extended to 3×3 systems of linear equations.

Cramer's Rule for 3 × 3 Systems

The solution to $\begin{cases} a_1x + b_1y + c_1z = k_1 \\ a_2x + b_2y + c_2z = k_2 \\ a_3x + b_3y + c_3z = k_3 \end{cases}$ is given by

$x = D_x/D$ and $y = D_y/D$ and $z = D_z/D$ where $D = \begin{vmatrix} a_1 & b_1 & c_1 \\ a_2 & b_2 & c_2 \\ a_3 & b_3 & c_3 \end{vmatrix}$ and $D \neq 0$,

$D_x = \begin{vmatrix} k_1 & b_1 & c_1 \\ k_2 & b_2 & c_2 \\ k_3 & b_3 & c_3 \end{vmatrix}$, $D_y = \begin{vmatrix} a_1 & k_1 & c_1 \\ a_2 & k_2 & c_2 \\ a_3 & k_3 & c_3 \end{vmatrix}$ and, $D_z = \begin{vmatrix} a_1 & b_1 & k_1 \\ a_2 & b_2 & k_2 \\ a_3 & b_3 & k_3 \end{vmatrix}$.

Solved Problem 7.4 Use Cramer's rule to solve the following:

$$\begin{cases} x+y+z=-1 \\ 2x-y+2z=4 \\ -x+2y-4z=-5 \end{cases}$$

Solution:

$$D=\begin{vmatrix} 1 & 1 & 1 \\ 2 & -1 & 2 \\ -1 & 2 & -4 \end{vmatrix}=1\begin{vmatrix} -1 & 2 \\ 2 & -4 \end{vmatrix}-1\begin{vmatrix} 2 & 2 \\ -1 & -4 \end{vmatrix}+1\begin{vmatrix} 2 & -1 \\ -1 & 2 \end{vmatrix}=9\neq 0$$

$$D_x=\begin{vmatrix} -1 & 1 & 1 \\ 4 & -1 & 2 \\ -5 & 2 & -4 \end{vmatrix}=-1\begin{vmatrix} -1 & 2 \\ 2 & -4 \end{vmatrix}-1\begin{vmatrix} 4 & 2 \\ -5 & -4 \end{vmatrix}+1\begin{vmatrix} 4 & -1 \\ -5 & 2 \end{vmatrix}=9$$

$$D_y=\begin{vmatrix} 1 & -1 & 1 \\ 2 & 4 & 2 \\ -1 & -5 & -4 \end{vmatrix}=1\begin{vmatrix} 4 & 2 \\ -5 & -4 \end{vmatrix}-(-1)\begin{vmatrix} 2 & 2 \\ -1 & -4 \end{vmatrix}+1\begin{vmatrix} 2 & 4 \\ -1 & -5 \end{vmatrix}=-18$$

$$D_z=\begin{vmatrix} 1 & 1 & -1 \\ 2 & -1 & 4 \\ -1 & 2 & -5 \end{vmatrix}=1\begin{vmatrix} -1 & 4 \\ 2 & -5 \end{vmatrix}-1\begin{vmatrix} 2 & 4 \\ -1 & -5 \end{vmatrix}+(-1)\begin{vmatrix} 2 & -1 \\ -1 & 2 \end{vmatrix}=0$$

If $D=0$, there is no unique solution to the system. The system is either inconsistent or dependent. If at least one of D_x, D_y, or D_z as well as D is zero, the system is inconsistent. If D, D_x, D_y, or D_z are all zero, the system is dependent.

Matrix Methods

We previously employed the addition (elimination) method to solve systems of linear equations. In that process, various operations were performed to alter the coefficients of variables in equations. Since our attention was primarily on the coefficients of the variables, we can streamline the process by writing these coefficients only in an orderly array. The array we shall employ is called a matrix. A *matrix* is a rectangular array of elements (entries). The *elements* are usually numbers or letters. The elements or entries are displayed in rows and columns within brackets or parentheses. The following illustrate the symbolism normally used.

$$\begin{bmatrix} 2 & 1 & 0 \\ -3 & 5 & -2 \\ 1 & 8 & 9 \end{bmatrix}, \begin{bmatrix} 1 & 2 & 4 \\ 3 & 5 & 6 \end{bmatrix}, \text{ and } \begin{bmatrix} 4 \\ -7 \end{bmatrix}$$

The *size*, *dimension*, or *order* of a matrix is specified by stating the number of rows followed by the number of columns. The size, dimension, or order of the above matrices is 3×3, 2×3, and 2×1, respectively. A *square matrix* has the same number of rows and columns.

We now illustrate how matrices are used to solve a system of linear equations. Consider the system

$$\begin{cases} 3s + t = 7 \\ 2s + 5t = 1 \end{cases}$$

The matrix $\begin{bmatrix} 3 & 1 \\ 2 & 5 \end{bmatrix}$ is called the *coefficient matrix* of the system. It simply consists of the coefficients of the variables in the equations. The matrix $\begin{bmatrix} 7 \\ 1 \end{bmatrix}$ is called the *constant matrix* of the system. It is comprised of the constants in the right members of the equations. The matrix

$$\begin{bmatrix} 3 & 1 & 7 \\ 2 & 5 & 1 \end{bmatrix}$$

is called the *augmented matrix* of the system. It consists of the coefficient matrix of the system with the constant matrix of the system annexed on the right.

Recall that equivalent equations have the same solution. *Equivalent systems of equations* likewise have the same solution(s). The associated augmented matrices of equivalent systems are *row-equivalent matrices.* Row-equivalent matrices are obtained by the following operations:

You Need to Know ✓

Elementary Row Operations

1. Interchange any two rows.
2. Multiply each element of any row by a nonzero constant.
3. Add a multiple of one row to another row.

Solved Problem 7.5 Perform the elementary row operations on the matrix:

$$\begin{bmatrix} 1 & 2 & 1 \\ 3 & 7 & 4 \end{bmatrix}$$

Solution: (a) Interchanging rows 1 and 2 yields

$$\begin{bmatrix} 3 & 7 & 4 \\ 1 & 2 & 1 \end{bmatrix}$$

(b) Multiplying each element of row 1 by 3 to obtain a new row 1 yields

$$\begin{bmatrix} 3 & 6 & 3 \\ 3 & 7 & 4 \end{bmatrix}$$

(c) Multiplying each element of row 1 by 3 and adding the resulting elements to the corresponding elements of row 2 to obtain a new row 2 yields

$$\begin{bmatrix} 1 & 2 & 1 \\ 0 & 13 & 7 \end{bmatrix}$$

Nonlinear System

A nonlinear equation has degree two or more. The graph is not a straight line; hence the term nonlinear. A *nonlinear system* has at least one nonlinear equation. Our concentration here is on solving nonlinear systems. The techniques employed eliminate variables, one by one, until an equation in one variable that we can solve remains. There are two methods used primarily: the substitution method and the addition method.

Substitution Method

The substitution method works very well when one of the equations in the system is linear. We simply solve the linear equation explicitly for one variable and substitute into the other equation. The resulting equation must be an equation we can solve.

Solved Problem 7.6 Solve the following system of equations using the substitution method:

$$\begin{cases} y = x^2 + 2x + 2 \\ x + y = 2 \end{cases}$$

Solution: Solve the linear equation for y, then substitute into the first equation and solve the result.

$$\begin{aligned} x + y = 2 &\Rightarrow y = 2 - x \\ 2 - x &= x^2 + 2x + 2 \\ 0 &= x^2 + 3x \\ 0 &= x(x+3) \Rightarrow x = 0 \text{ or } x = -3 \end{aligned}$$

If $x = 0$, $y = 2 - x = 2 - 0 = 2$. If $x = -3$, $y = 2 - x = 2 - (-3) = 5$. The solution set is $\{(0, 2), (-3, 5)\}$. Both ordered pairs satisfy the original equations.

Addition Method

If both equations in a system are second-degree in both variables, the addition (elimination) method often works well.

Solved Problem 7.7 Solve the following system of equations using the addition method:

$$\begin{cases} s^2 - t^2 = -5 \\ 4s^2 + t^2 = 25 \end{cases}$$

Solution: Adding the two equations yields an equation in one unknown:

$$5s^2 = 20 \Rightarrow s = \pm\sqrt{4} = \pm 2$$

Now use back substitution to find t. We shall substitute into $4s^2 + t^2 = 25$. If $s = 2$,

$$\begin{aligned} 4s^2 + t^2 &= 25 \\ 4(2)^2 + t^2 &= 25 \\ 16 + t^2 &= 25 \\ t^2 &= 9 \\ t &= \pm\sqrt{9} = \pm 3 \end{aligned}$$

If $s = -2$,

$$\begin{aligned} 4s^2 + t^2 &= 25 \\ 4(-2)^2 + t^2 &= 25 \\ 16 + t^2 &= 25 \\ t^2 &= 9 \\ t &= \pm\sqrt{9} = \pm 3 \end{aligned}$$

The solution set is $\{(2, 3), (2, -3), (-2, 3), (-2, -3)\}$. It can be verified that all four ordered pairs satisfy both equations.

Systems of Inequalities

Systems of inequalities can be solved graphically. The solution set of a system of inequalities is simply the intersection of the solution sets of the individual inequalities. We first graph each inequality separately. Finally, the graphs are combined to display the solution set of the system.

Solved Problem 7.8 Solve the following system of equations graphically:

$$\begin{cases} y \geq 2x - 2 \\ y \leq -3x + 2 \end{cases}$$

Solution: The graph of $y \geq 2x - 2$ is found by graphing $y = 2x - 2$, then choosing a test point, such as (0, 0), and finally shading the appropriate region. The graph of $y = 2x - 2$ is a line with slope 2 and y-intercept −2 (see Figure 7-4 (*a*)). The graph of $y \leq -3x + 2$ is done similarly (see Figure 7-4 (*b*)). Determine the intersection as shown in Figure 7-4 (*c*).

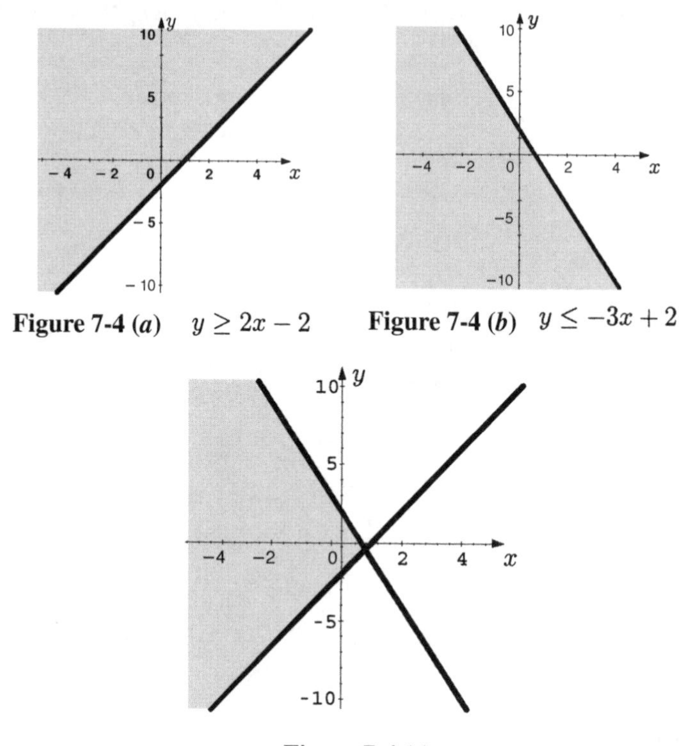

Figure 7-4 (*a*) $y \geq 2x - 2$

Figure 7-4 (*b*) $y \leq -3x + 2$

Figure 7-4 (*c*)

We choose an arbitrary point in the shaded region, such as (−2, 0), to verify that it satisfies each inequality in the system.

Chapter 8
RELATIONS AND FUNCTIONS

IN THIS CHAPTER:

- ✔ *Basic Concepts*
- ✔ *Function Notation and the Algebra of Functions*
- ✔ *Distance and Slope Formulas*
- ✔ *Linear Equation Forms*
- ✔ *Types of Functions*
- ✔ *Inverse Relations and Functions*

Basic Concepts

The *Cartesian Product* is basic to the mathematical concepts of relations and functions.

Definition 1. $A \times B = \{(a, b) | a \in A \text{ and } b \in B\}$ is called the *Cartesian Product*.

The *Cartesian Product* of two sets is the set of all ordered pairs for which the first element is a member of the first set and the second element is a member of the second set.

Definition 2. A *relation* is any set of ordered pairs. The set of first elements is the *domain* and the set of second elements is the *range*.

A *relation* of real numbers is a subset of $R \times R$ (where R symbolizes the set of real numbers). A *graph* is a relation displayed on a coordinate system.

Definition 3. A *function* is a set of ordered pairs or relation for which no two distinct ordered pairs have the same first element.

Definition 3. **(Alternate)** A *function* from a set A to a set B is a mapping, rule, or correspondence that assigns to each element x in set A exactly one element y in set B. (Each x is mapped to a unique y.)

A function is a relation with an added condition. The added condition is that for a function, no x value can have more than one y value associated with it. To determine whether a relation in graphical format is a function, we can use the *Vertical Line Test*.

Vertical Line Test: A graph is the graph of a function if no vertical line intersects the graph in more than one point.

If any vertical line intersects a graph in two or more points, then the graph is not the graph of a function. The reason is this. Two points on the same vertical line indicate two different ordered pairs that have the same first element. Thus, it can't be a function.

A function is frequently described using set builder notation as in $\{(x, y) | y = 3x - 5\}$. In this form, the first variable in the ordered pair is called the *independent variable* and the second variable is the *dependent variable* (since its value depends on the value of the first variable).

The concepts of domain and range apply to functions since all functions are also relations though not all relations are functions. When the function is described using an equation and the domain isn't stated (given explicitly), then we need to find the *implicit domain*. The implicit domain is the set of all values for the independent variable that result in real number values for the dependent variable. For the functions we will consider in this chapter, the implicit domain is found by excluding (from the set of real numbers) any values that result in (1) division by zero or (2) square roots of negative numbers.

Function Notation and the Algebra of Functions

The definition of a function indicates that it is a set of ordered pairs. To distinguish between two or more functions, we can give each a name, as in $f = \{(x, y) | y = x^2 + 3x\}$ and $g = \{(x, y) | y = x / (x - 2)\}$. In addition, to be more efficient, a more compact notation has been created to designate functions. The function f above would be designated as $f(x) = x^2 + 3x$, and g would be $g(x) = x / (x - 2)$. This notation does not imply multiplication even though it looks like it should. This is simply new notation. One advantage of this notation is its simplicity: e.g., $f(x)$ above could be used anywhere $x^2 + 3x$ would be used. The independent variable x in this notation is just a place-holder for a number or expression.

Definition 4. For all x common to the domains of f and g, the arithmetic of f and g is given by the following:

1. The *sum* is $(f + g)(x) = f(x) + g(x)$
2. The *difference* is $(f - g)(x) = f(x) - g(x)$
3. The *product* is $(fg)(x) = f(x) \cdot g(x)$
4. The *quotient* is $\left(\frac{f}{g}\right)(x) = \frac{f(x)}{g(x)}$, for $g(x) \neq 0$

Definition 5. For all x for which the expression is defined, the *composition of functions*, symbolized as f, is defined $(f \circ g)(x)$ to be $f[g(x)]$.

The composition of functions, $(f \circ g)(x) = f[g(x)]$, evaluates the function f at $g(x)$. It requires the value for $g(x)$ to be in the domain of f.

Distance and Slope Formulas

Distance Between Points in the Plane

Basic to the distance formula is the following theorem:

Pythagorean Theorem: The sum of the squares of the lengths of the legs in a right triangle (Figure 8-1) is equal to the square of the length of the hypotenuse or

$$(leg_1)^2 + (leg_2)^2 = (hypotenuse)^2$$

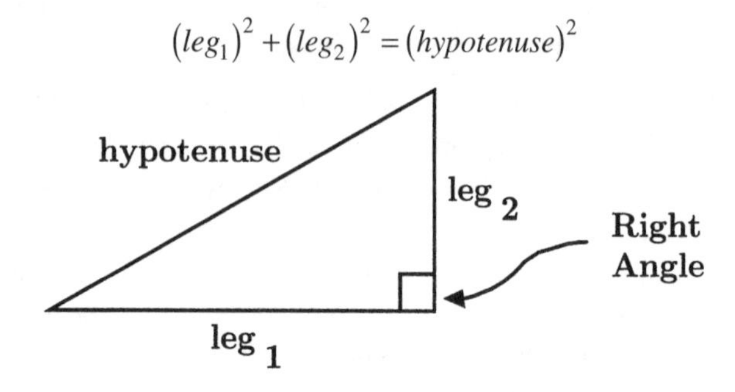

Figure 8-1

The distance formula can be better understood using the Pythagorean Theorem. Figure 8-2 below graphically shows the development of the formula for the distance between points (1, –2) and (5, 4), that is, the distance between the endpoints of the bold line segment d in the figure.

By the Pythagorean Theorem, we can now see that $d^2 = 4^2 + 6^2$ or $d = \sqrt{4^2 + 6^2} = \sqrt{16 + 36} = \sqrt{52} = 2\sqrt{13} \approx 7.2111$.

Distance Formula: The distance d between (x_1,y_1) and (x_2,y_2) is

$$d = \sqrt{(x_2 - x_1)^2 + (y_2 - y_1)^2}$$

The distance between any two points in the plane is the square root of the sum of the squares of the difference of the x-coordinates and the difference of the y-coordinates.

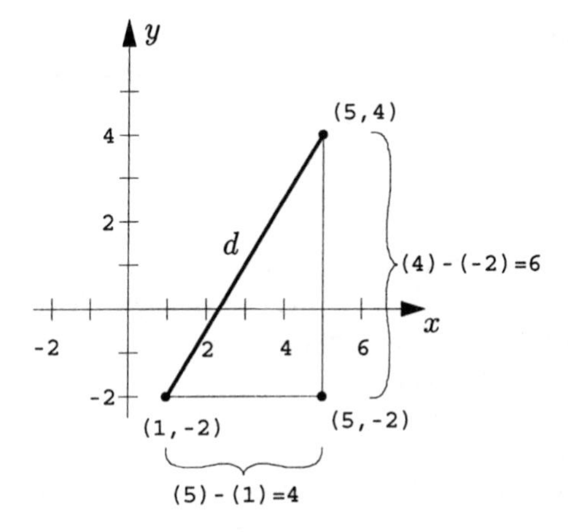

Figure 8-2

Slope of a Line

A nonvertical line that passes through two distinct points in the plane has an inclination to the horizontal. We call that inclination the slope. It is calculated by dividing the difference in the y-coordinates of the two points by the difference in their x-coordinates. Figure 8-3 indicates the slope of the bold line segment is $slope = \frac{6}{4} = 1.5$.

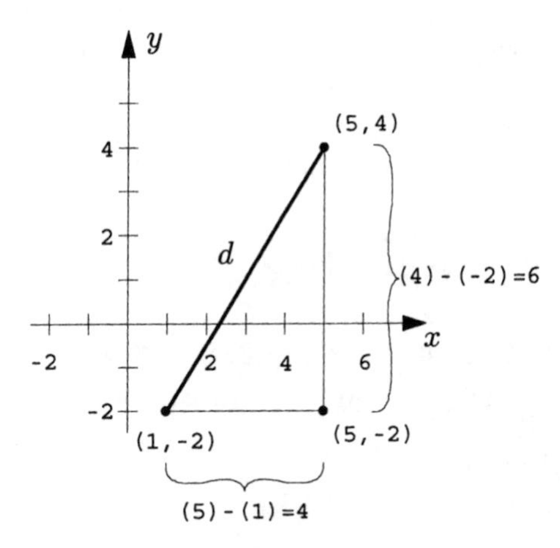

Figure 8-3

Slope Formula: The slope of the line passing through the points (x_1,y_1) and (x_2,y_2) is symbolized by the letter m and defined to be:

$$m = \frac{y_2 - y_1}{x_2 - x_1}$$

if the denominator is not zero. If the denominator is zero, the slope is undefined, the line has no slope associated with it. In other words, the slope of a line passing through two points is the directed change in y divided by the directed change in x.

Solved Problem 8.1 Find the slope of the line passing through each of the following pairs of points: (a) (2, 4) and (5, 10) and (b) (1, –2) and (–3, –5).

Solution:

(a) $m = \frac{y_2 - y_1}{x_2 - x_1} = \frac{10-4}{5-2} = \frac{6}{3} = 2$

(b) $m = \frac{y_2 - y_1}{x_2 - x_1} = \frac{(-5)-(-2)}{(-3)-(1)} = \frac{-3}{-4} = \frac{3}{4}$

Linear Equation Forms

From the formula for the slope of a line, we can develop formulas for equations of a line provided we are given enough information about the line. Using Figure 8-4 below, we can find an equation of the line passing through the two points (–2, 3) and (6, –1) by identifying a relation between the coordinates of the general point (x, y).

> **Remember**
>
> In general, to find an equation of a line, we need to know two things: its slope and a point on the line.

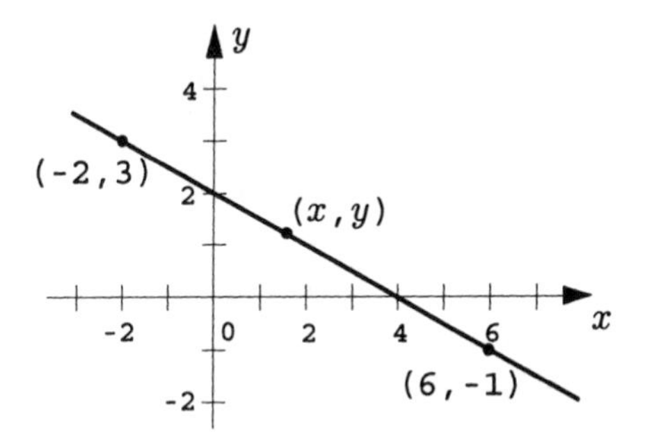

Figure 8-4

Definition 6. For any nonvertical line, the *Point-Slope Form* of the equation of a line is

$$y - y_1 = m(x - x_1)$$

where (x_1, y_1) is a known point on the line and m is the slope of the line.

In the point-slope form of the equation of a line, we use the known point and known slope to write an equation that identifies every point on the line.

Definition 7. The *Slope-Intercept Form* of the equation of a line is

$$y = mx + b$$

The slope-intercept form of a line requires y to be isolated on one side of the equation. In that form, the coefficient of x is the slope and the constant term is the y-intercept. The student should be careful to note that the coefficient of x is the slope of the line in this form only—the coefficient of x is not the slope in the other forms of the equation of the line.

Solved Problem 8.2 Determine the slope and y-intercept of the line given by the equation $3x - 5y = 15$.

Solution: We first solve the equation for y:

$$
\begin{aligned}
3x - 5y &= 15 \\
-5y &= -3x + 15 \\
y &= \frac{-3x + 15}{-5} \\
y &= \frac{3}{5}x - 3
\end{aligned}
$$

Now the slope can be read from the coefficient of x as $\frac{3}{5}$ and the y-intercept is read from the constant as -3.

Definition 8. The *General Form* of the equation of a line is $ax + by + c = 0$; the *Standard Form* is $ax + by = c$. (Traditionally, a is nonnegative and all fractions are cleared.)

Both the general form and the standard form of the equation of the line require that either a or b is nonzero. Otherwise, the equation becomes $c = 0$. If c isn't zero, then this is a contradiction; if c is zero, then it is an identity. As you can see, $x + y + 3 = 0$ and $2x + 2y + 6 = 0$ both represent the same line since they are equivalent equations (multiply both sides of the first equation by 2 to obtain the second equation). Therefore, the solutions in the general and standard forms that you find may differ from the solutions we give, but your solutions should be equivalent to ours.

Definition 9. The *Intercept Form* of the equation of a line is $x/r + y/s = 1$ where r is the x-intercept of the line and s is the y-intercept. Neither r nor s may be zero so this form cannot be used when the line passes through the origin.

The Intercept Form is a variation of the Standard Form requiring $c = 1$.

Parallel Lines. Nonvertical *parallel* lines have equal slopes. Lines with equal slopes are parallel.

If line l_1 with slope m_1 is parallel to line l_2 with slope m_2, then $m_1 = m_2$. Conversely, if $m_1 = m_2$, then l_1 is parallel to l_2.

Perpendicular Lines. Two *perpendicular* lines, neither of which is vertical, have slopes whose product is −1. (Alternatively, nonvertical perpendicular lines have slopes that are negative reciprocals of one another.) Conversely, two lines having slopes whose product is −1 are perpendicular.

If line l_1 with slope m_1 is perpendicular to line l_2 with slope m_2, then $m_1 \cdot m_2 = -1$ or equivalently $m_2 = -1/m_1$. In the reverse perspective, if $m_1 \cdot m_2 = -1$, then l_1 is perpendicular to l_2.

Types of Functions

The graph of a function f is normally the graph of the equation $y = f(x)$. We will use this equation to identify the graphs of the following special types of functions.

Definition 10. Any function of the form $f(x) = ax + b$ is identified as a *linear function.*

Since $y = f(x) = ax + b$, we see that the graph of a linear function is a straight line with slope $m = a$ and y-intercept $= b$. All linear functions graph into nonvertical lines and all nonvertical lines can be represented by linear functions.

Solved Problem 8.3 Graph $f(x) = -3x + 2$.

Solution: We first create a table of points or ordered pairs to plot using $f(x) = y = -3x + 2$ and then graph the line through the points. See Figure 8-5.

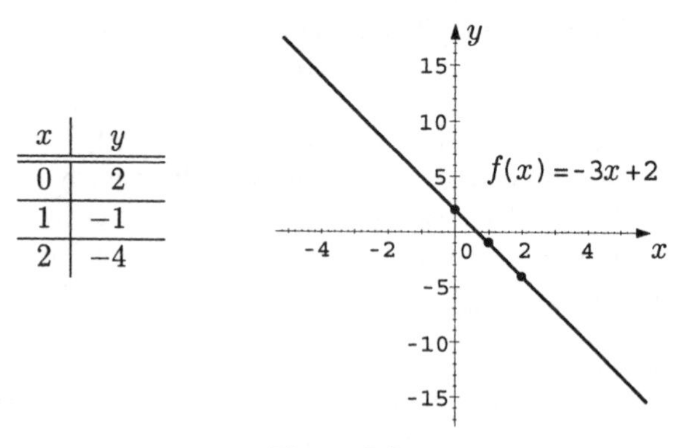

x	y
0	2
1	−1
2	−4

Figure 8-5

Definition 11. Any function of the form $f(x) = ax^2 + bx + c$ where $a \neq 0$ is identified as a *quadratic function*.

Since $y = f(x) = ax^2 + bx + c = a\left(x + \dfrac{b}{2a}\right)^2 + \dfrac{4ac - b^2}{4a}$, we know that the graph of a quadratic function is a parabola. The graph opens upward if $a > 0$, downward if $a < 0$, and has a vertex at

$$\left(\frac{-b}{2a}, f\left(\frac{-b}{2a}\right)\right) = \left(\frac{-b}{2a}, \frac{4ac - b^2}{4a}\right)$$

Every quadratic function graphs into a parabola with a vertical axis of symmetry and every parabola with a vertical axis can be represented by a quadratic function.

Both linear and quadratic functions are examples of a more general type that we will not analyze in this text. That type is the *polynomial function*. A polynomial function is of the form

$$f(x) = a_n x^n + a_{n-1} x^{n-1} + \cdots + a_2 x^2 + a_1 x + a_0$$

where n is a nonnegative integer and $a_n \neq 0$ (unless $n = 0$, meaning $f(x) = a_0$ and $f(x) = 0$ is valid).

Definition 12. The function $f(x) = \sqrt{x}$, where x is a real number and $x \geq 0$, is identified as a *square root function.*

The square root function is the top half of a parabola opening to the right with vertex at the origin. See Figure 8-6.

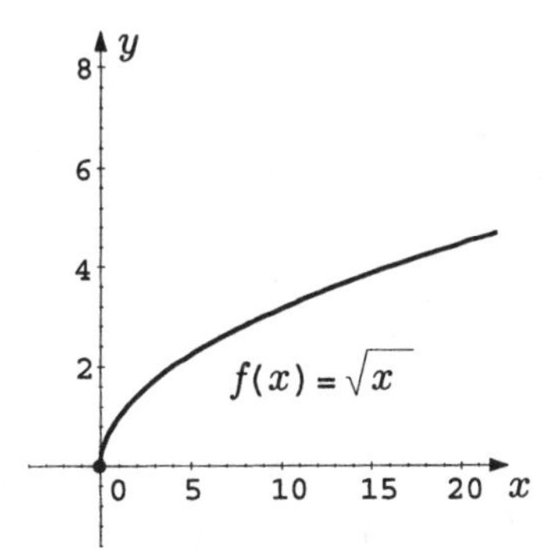

Figure 8-6

Definition 13. The function $f(x) = |x|$ is identified as the *absolute value function.*

At the origin, the graph has a sharp turn or corner, which is called the vertex. The graph resembles the letter "V." As with the square root function, variations exist for the absolute value function and are of the form $f(x) = a|x + b| + c$. The vertex is at the point $(-b, c)$ and the graph opens upward (in a "V") if $a > 0$, or downward (in an inverted "V") if $a < 0$.

Inverse Relations and Functions

Definition 14. The *inverse* of a relation or function is the relation that results from interchanging the x- and y-coordinates of all ordered pairs.

Solved Problem 8.4 Determine the inverse of $f(x) = -2x + 5$.

Solution: This is the function, $f = \{(x, y)|y = -2x + 5\}$. We can interchange the x- and y-coordinates to obtain the inverse by either switching the order in the ordered pair directly, as (y, x), or by leaving the ordered pair alone and switching its "order" indirectly by interchanging x and y in the equation: $x = -2y + 5$. The latter method is normally used. So, for this function, $f^{-1} = \{(x, y)|x = -2y + 5\}$.

You Need to Know 

The process generally used to find the inverse of a function $f(x)$, uses three steps:

1. Set $y = f(x)$.
2. Interchange x and y. That is, replace each occurrence of x with y and each occurrence of y with x.
3. Solve the resulting equation for y.

Chapter 9
Exponential and Logarithmic Functions

In This Chapter:

- ✔ *Exponential Functions*
- ✔ *Logarithmic Functions*
- ✔ *Properties of Logarithms*
- ✔ *Exponential and Logarithmic Equations*
- ✔ *Applications*

Exponential Functions

Definition 1. An *exponential function* is of the form $f(x) = b^x$ where $b > 0$, $b \neq 1$, and x is any real number.

The domain of the exponential function is all real numbers and its range is all positive real numbers (independent of the value of b); b^x cannot be either negative or zero for $b > 0$.

Solved Problem 9.1 Graph each of the following exponential functions: (a) $f(x) = 2^x$ and (b) $g(x) = 10^x$.

Solution:

(a)

x	2^x
-3	$\frac{1}{8}$
-2	$\frac{1}{4}$
-1	$\frac{1}{2}$
0	1
1	2
2	4
3	8

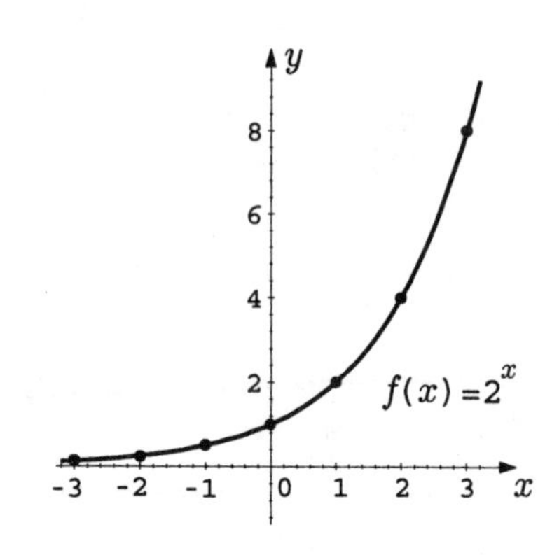

(b)

x	10^x
-2	$\frac{1}{100}$
-1	$\frac{1}{10}$
0	1
1	10
2	100

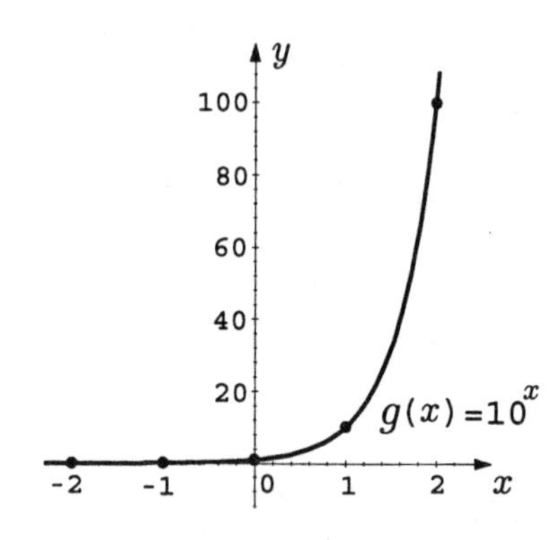

In general, if $b > 1$ the exponential function $f(x) = b^x$ increases in value as x increases, while if $0 < b < 1$, the function decreases in value as x increases. Because of this and because of their applications, exponential functions are frequently called growth or decay functions, respectively. A special exponential function, $f(x) = e^x$, is used in many applications. It is based on an irrational number symbolized by "e." We call e the *base of the natural exponential function*. To 15 decimal places,

$$e \approx 2.718281828459045$$

Solved Problem 9.2 Graph $f(x) = e^x$.

Solution:

$y = f(x) = e^x \Rightarrow$

x	y
-3	4.9787×10^{-2}
-2	0.13534
-1	0.36788
0	1
1	2.7183
2	7.3891
3	20.086

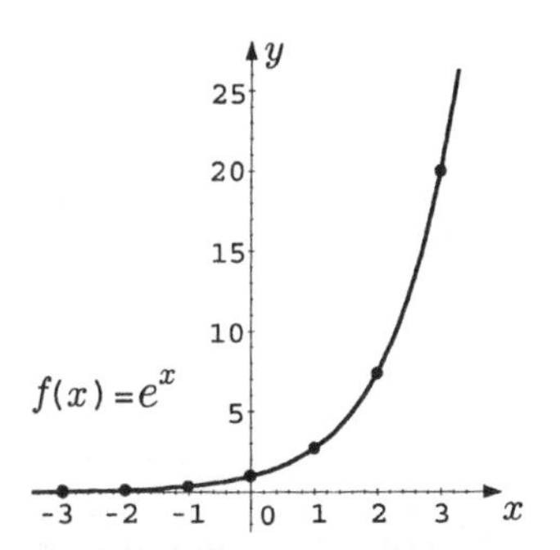

You Need to Know 

Exponential Properties

For a, $b > 0$ and a, $b \neq 1$,

1. $a^n = b^n$ if and only if $a = b$.
2. $a^n = a^m$ if and only if $n = m$.

Logarithmic Functions

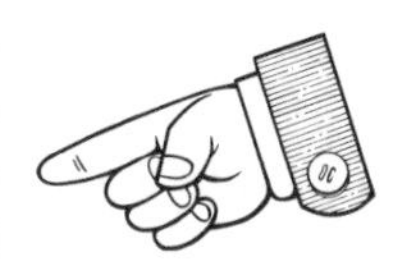

Since the exponential function is a one-to-one function, we know its inverse is also a function. The inverse of the exponential function is the logarithmic function, or log for short. We attempt to find the inverse of $f(x) = b^x$ by the conventional method: (1) $f(x) = y = b^x$; (2) $x = b^y$; and (3) then solve for y. To solve for y, however, requires a new function, namely, $\log_b x$. This function is read: "the logarithm to the base b of x." The inverse then is $f^{-1}(x) = y = \log_b x$.

Definition 2. If b and x are positive real numbers with $b \neq 1$, the function

$$f(x) = \log_b x$$

is called the *logarithmic function* to the base b.

The domain of the logarithmic function is all positive real numbers (the range of the exponential function) and its range is all real numbers (the domain of the exponential function).

Because the exponential and logarithmic functions are inverses of one another, the following is identified as the fundamental relationship between them.

Remember

$b^a = c$ is equivalent to $a = \log_b c$

Note that b is the base in each equation; the base of the power in the exponential equation and the base of the logarithm in the logarithmic equation. The value of a logarithm is, in essence, an exponent. In the equivalence stated above, it is the exponent to which the base b must be raised to obtain the number c. This relationship is the key relationship employed in many problems that involve the exponential and logarithmic functions.

Solved Problem 9.3 Graph each of the following: (a) $y = \log_2 x$; (b) $y = \log_5 x$; (c) $y = \log_{0.1} x$.

Solution:

(a) $y = \log_2 x$

y	$x = 2^y$
-2	$\frac{1}{4}$
-1	$\frac{1}{2}$
0	1
1	2
2	4
3	8

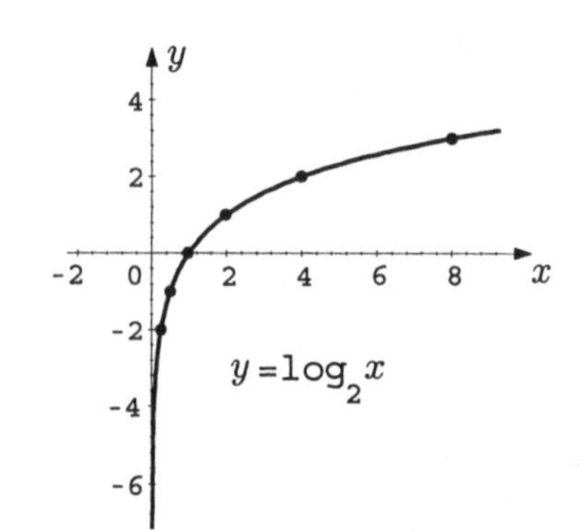

(b) $y = \log_5 x$

y	$x = 5^y$
-2	$\frac{1}{25}$
-1	$\frac{1}{5}$
0	1
1	5
2	25

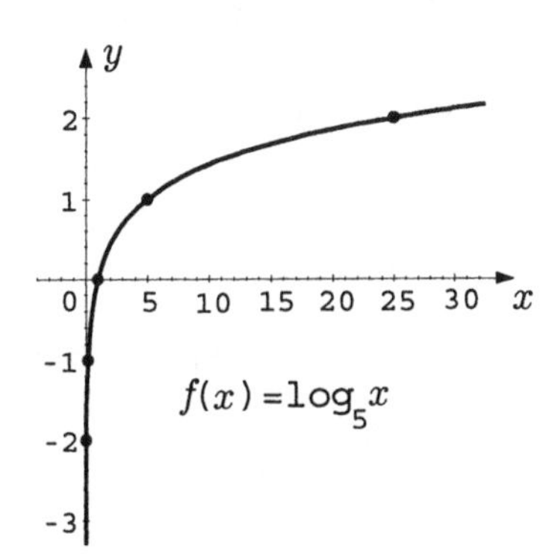

(c) $y = \log_{0.1} x$

y	$x = 0.1^y$
-2	100
-1	10
0	1
1	0.1
2	0.01

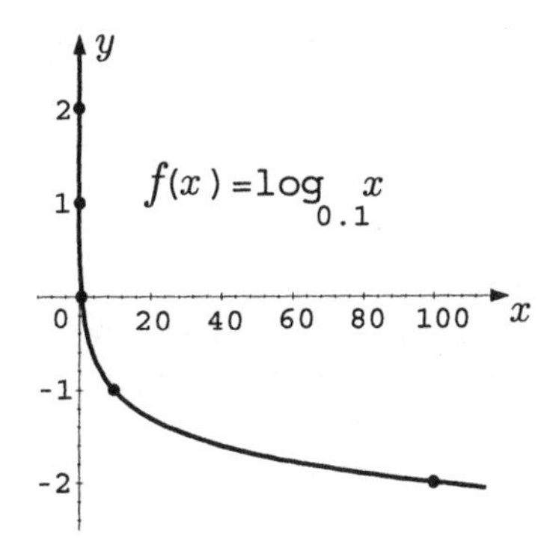

The following identities frequently prove useful:

Identities

$$\log_b b^x = x$$

$$b^{\log_b x} = x$$

$$\log_b 1 = 0$$

Common Logs and Natural Logs

The following abbreviations allow us to represent the indicated logarithms without specifying a base:

Common logarithms: $\log_{10} x = \log x$

Natural logarithms: $\log_e x = \ln x$

Properties of Logarithms

The following five basic properties are used for manipulations involving logarithms.

Basic Logarithmic Properties

1. $\log_b(ac) = \log_b a + \log_b c$
2. $\log_b\left(\frac{a}{c}\right) = \log_b a - \log_b c$
3. $\log_b\left(a^c\right) = c\log_b a$
4. $\log_b a = \frac{\log_c a}{\log_c b}$
5. $\log_b a = \log_b c$ if and only if $a = c$

In all five properties above, $a > 0$, $b > 0$, $b \neq 1$, and $c > 0$. In property 4, $c \neq 1$. Property 4 above is frequently referred to as the "change of base formula." It is used to change the base of a logarithm from a base that cannot be evaluated directly using a calculator into a base that can (using a quotient of logarithms).

Exponential and Logarithmic Equations

Equations involving exponential functions are called *exponential equations*. Equations involving logarithmic functions are called *logarithmic equations*. Both frequently require the use of logarithmic properties. When introducing logarithms, generally we want to use either natural logarithms (ln) or common logarithms (log).

Solved Problem 9.4 Solve each of the following for x: (a) $2^x = 15$; (b) $12^{2-x} = 20$; and (c) $\log_3 x + \log_3 4 = 2$.

Solution:

(a) $2^x = 15$

$$\begin{aligned} \ln\left(2^x\right) &= \ln 15 \\ x \ln 2 &= \ln 15 \\ x &= \frac{\ln 15}{\ln 2} \\ x &\approx \frac{2.708050}{0.6931472} \\ x &\approx 3.90689 \end{aligned}$$

(b) $12^{2-x} = 20$

$$\begin{aligned} \ln\left(12^{2-x}\right) &= \ln 20 \\ (2-x)\ln 12 &= \ln 20 \\ 2\ln 12 - x\ln 12 &= \ln 20 \\ -x\ln 12 &= \ln 20 - 2\ln 12 \\ x &= \frac{\ln 20 - 2\ln 12}{-\ln 12} \\ x &\approx \frac{-1.974081}{-2.484907} \\ x &\approx 0.794429 \end{aligned}$$

(c) $\log_3 x + \log_3 4 = 2$

$$\begin{aligned} \log_3 x + \log_3 4 &= 2 \\ \log_3(4x) &= 2 \\ 4x &= 3^2 \\ x &= \frac{9}{4} = 2.25 \end{aligned}$$

Applications

Interest

Compound interest occurs when an initial amount of money earns interest at a constant rate at the end of each period for multiple periods. If that interest rate per period is i and the initial amount of money or principal is P, then the accumulated amount at the end of each period A is

$$A = P + Pi = P(1+i)$$
$$A = P(1+i) + P(1+i)i = P(1+i)(1+i) = P(1+i)^2$$
$$A = P(1+i)^2 + P(1+i)^2 i = P(1+i)^2(1+i) = P(1+i)^3$$
$$\vdots$$
$$A = P(1+i)^n$$

Solved Problem 9.5 Determine the time required for the principal to double in value when it is compounded quarterly at an annual rate of 8%.

Solution: From the information given in the problem,

$i = 0.08/4$, $n = 4t$ where t is the number of years required. Then,

$$A = 2P = P\left(1 + \frac{0.08}{4}\right)^{4t}$$
$$2 = \left(1 + \frac{0.08}{4}\right)^{4t}$$
$$\ln 2 = \ln\left(1 + \frac{0.08}{4}\right)^{4t} = 4t \ln\left(1 + \frac{0.08}{4}\right)$$
$$t = \frac{\ln 2}{4 \ln\left(1 + \frac{0.08}{4}\right)} \approx 8.7507 \text{ years}$$

The principal will double in value in just over 8.75 years.

For a principal to be compounded continuously (instead of periodically), the formula is $A = Pe^{rt}$ where A is the accumulated amount, P is the principal, r is the annual interest rate and t is the number of years of compounding.

Exponential Growth and Decay

Exponential growth and decay functions follow the equation $A = A_0e^{kt}$ where A is the final amount, A_0 is the initial amount, t is the time span involved, and k is related to the rate of growth or decay. k is positive for growth functions and k is negative for decay functions.

Solved Problem 9.6 What is the percentage growth rate per hour for a bacteria culture that grew from 500,000 twelve hours ago to 10,000,000 now?

Solution: The original number of bacteria was $A_0 = 500{,}000$ and now is $A = 10{,}000{,}000$. To find k, the constant for the problem, solve the following:

$$\begin{aligned}
10{,}000{,}000 &= 500{,}000e^{k(12)} \\
\frac{10{,}000{,}000}{500{,}000} &= e^{k(12)} \\
20 &= e^{12k} \\
12k &= \ln 20 \\
k &= \frac{\ln 20}{12} \\
k &= 0.249644
\end{aligned}$$

Now we can use this value for k to determine the number of bacteria present after 1 hour.

$$\begin{aligned}
A &= 500{,}000e^{k(1)} \\
A &= 500{,}000e^{(0.249644)(1)} \\
A &= 500{,}000(1.28357) \\
A &= 641{,}785
\end{aligned}$$

By this, we can determine the growth rate for that first hour (and therefore for each hour) to be

$$\frac{641,785 - 500,000}{500,000} = 0.28357$$

The percentage growth rate from 500,000 to 10,000,000 in 12 hours is 28.357% per hour.

Chapter 10

SEQUENCES, SERIES, AND THE BINOMIAL THEOREM

IN THIS CHAPTER:

- ✔ *Sequences*
- ✔ *Series*
- ✔ *The Binomial Theorem*

Sequences

A *sequence* is an ordered list of numbers. The numbers 3, 6, 9, 12 form one sequence, while 3, 9, 6, 12 form another. The stated sequences are different since their order is different. The *terms* of a sequence consist of the expressions separated by commas. The stated sequences have four terms. A *finite sequence* has a last term while an *infinite sequence* has *no* last term. The ordered list 3, 6, 9, 12, … is an infinite sequence. The ellipsis symbol, "…", means the terms of the sequence continue in the pattern indicated without end. There are infinitely many terms in the list. An ellipsis is sometimes used to represent a finite number of ordered terms also.

The terms of a sequence often follow a particular pattern. In those instances, we can determine the general term that expresses every term of the sequence. For example,

Sequence	**General Term**
3, 6, 9, 12, …	$3n$
1, 3, 5, 7, …	$2n - 1$
2, 4, 8, 16, …	2^n

The variable n represents a positive integer. The first term of the sequence is obtained when $n = 1$, the second term is obtained when $n = 2$, and so on. The general term of a sequence specifies a function that produces the sequence when evaluated at the natural numbers. In other words, a sequence is a function whose domain is the set of natural numbers and range is some subset of real numbers.

It is customary to use a_n to represent the nth term or general term of a sequence. Thus, a_1 is the first term, a_2 is the second term, and so on. The entire sequence 3, 6, 9, 12, … is represented by $a_n = 3n$. We simply replace n by 1, 2, 3, … in $3n$ to obtain the successive terms of the sequence.

An *arithmetic sequence* or *arithmetic progression* is a sequence such that successive terms differ by the same constant. The constant difference is represented by d and is given by $d = a_{i+1} - a_i$ for all positive integers i. The sequence 2, 5, 8, 11, … is an arithmetic sequence. Successive terms differ by 3, that is $d = 3$.

You Need to Know ✓

There is an explicit formula for the nth term of an arithmetic sequence in general. The nth term of an arithmetic sequence with common difference d is $a_n = a_1 + (n - 1)\, d$.

A *geometric sequence* or *geometric progression* is a sequence such that each successive term is obtained by multiplying a constant times the previous term. Equivalently, the quotient (ratio) of successive terms is the same constant. This constant is called the *common ratio*, and is represented by r. The common ratio is given by $r = a_{i+1}/a_i$ for all positive integers i and $r \neq 0$. The sequence 3, 9, 27, 81, ... is a geometric sequence with common ratio $r = 3$.

The formula for the *nth* term of a geometric sequence with common ratio r is given below:

$$a_n = a_1 r^{n-1}$$

Solved Problem 10.1 Find the *nth* term, a_n, and the *10th* term, a_{10}, of the following geometric sequence: 5, 10, 20, ...

Solution: We first must find the common ratio $r = \frac{a_2}{a_1} = \frac{10}{5} = 2$. Therefore, $a_n = a_1 r^{n-1} = 5 \cdot 2^{n-1}$. Thus, $a_{10} = 5 \cdot 2^{10-1} = 5 \cdot 2^9 = 2{,}560$.

Series

Consider the arithmetic sequence given by 2, 5, 8, ..., $3n - 1$, ... We add successive terms to generate a sequence of partial sums. We employ S_n to represent partial sums. The *nth* partial sum of an arithmetic sequence, S_n, is the sum of the first n terms of the sequence.

$$\begin{aligned}
S_1 &= 2 \\
S_2 &= 2 + 5 = 7 \\
S_3 &= 2 + 5 + 8 = 15 \\
S_4 &= 2 + 5 + 8 + 11 = 26 \\
&\vdots \\
S_n &= 2 + 5 + 8 + \cdots + 3n - 1
\end{aligned}$$

There is a formula for the *nth* partial sum of an arithmetic sequence, given by:

$$S_n = \frac{n}{2}(a_1 + a_n)$$

If a_n is replaced by its equivalent $a_1 + (n - 1)\,d$, an alternate formula is obtained:

$$S_n = \frac{n[2a_1 + (n-1)d]}{2}$$

The arithmetic sequence 1, 2, 3, ..., n is a sequence with common difference $d = 1$, $a_1 = 1$, and $a_n = n$. The sequence is simply the first n positive integers. A formula for the sum of the first n positive integers or counting numbers can be found using the formula for the *nth* partial sum of an arithmetic sequence. It is

$$S_n = \left(\frac{n}{2}\right)(a_1 + a_n) = \left(\frac{n}{2}\right)(1+n) = \frac{[n(n+1)]}{2}$$

This useful result is stated below. The sum of the first n positive integers is:

$$S_n = \frac{n(n+1)}{2}$$

There is a convenient notation that is used for partial sums. It is called the *summation notation*. The sum of the first n terms of a sequence a_n is represented by:

$$S_n = \sum_{i=1}^{n} a_i = a_1 + a_2 + a_3 + \cdots + a_n$$

The letter i (other letters are employed also) is called the *index of the summation*; 1 is the *lower limit of the summation*; n is the *upper limit of the summation*. The "Σ" symbol is the capital Greek letter sigma; it tells us to find a sum. The sum of the terms of a finite sequence is called a *finite series*.

The notation above means to find the sum beginning with the term obtained when $i = 1$. We then increase i by one each time to obtain the successive terms in the sum until $i = n$ for the last term. The terms obtained are then added.

It is possible to find a general formula for the *nth* partial sum of a geometric sequence also from the following:

$$S_n = \frac{a_1(1-r^n)}{1-r} = \frac{a_1(r^n-1)}{r-1}, r \neq 1$$

Use the latter form if $r > 1$ in order to avoid negative numerators and denominators.

Consider the geometric sequence $2, \frac{2}{3}, \frac{2}{3^2}, \frac{2}{3^3}, \ldots, \frac{2}{3^{n-1}}, \ldots$. By inspection, we observe that $r = \frac{1}{3}$. We wish to find the sum of *all* of the terms in the sequence. Is it possible to add infinitely many terms in the sequence?

The sum of the terms of an infinite geometric sequence is called an *infinite geometric series*. In general, the sum of all the terms in an infinite sequence, geometric or not, is called an *infinite series*.

The formula for the *nth* partial sum of a geometric sequence was given as $S_n = \frac{a_1(1-r^n)}{1-r}$. If a_n is a geometric sequence with first term a_1 and $|r| < 1$, the sum of all the terms S is given by $S = \frac{a_1}{1-r}$.

The Binomial Theorem

There are circumstances in mathematics in which $(a + b)^n$ is written as the sum of its terms. The process employed is called *expanding the binomial* or *writing the binomial in expanded form*. We now apply the special product forms as well as the distributive property to obtain powers of $a + b$ for various n. We are searching for patterns that will be helpful in the future. The following array is obtained after a certain amount of effort.

$n = 0$	$(a+b)^0$	1
$n = 1$	$(a+b)^1$	$a + b$
$n = 2$	$(a+b)^2$	$a^2 + 2ab + b^2$
$n = 3$	$(a+b)^3$	$a^3 + 3a^2b + 3ab^2 + b^3$
$n = 4$	$(a+b)^4$	$a^4 + 4a^3b + 6a^2b^2 + 4ab^3 + b^4$
$n = 5$	$(a+b)^5$	$a^5 + 5a^4b + 10a^3b^2 + 10a^2b^3 + 5ab^4 + b^5$

Observe the variable parts in each expansion for $n = 1, 2, 3, 4$, and 5.

1. The first term is a^n. The exponent on a decreases by 1 in successive terms.

2. The exponent on b increases by 1 in successive terms. The last term is b^n.

3. The sum of the exponents in each term is n.

Now take note of the numerical coefficients in each expansion. The following array of coefficients is obtained by omitting the variable factors in each term.

$n=0$						1					
$n=1$					1		1				
$n=2$				1		2		1			
$n=3$			1		3		3		1		
$n=4$		1		4		6		4		1	
$n=5$	1		5		10		10		5		1

The triangular array displayed above is called *Pascal's Triangle*. It is named in honor of Blaise Pascal, a 17th century mathematician and philosopher.

The following patterns in the triangular array can be identified:

1. The first and last coefficient in each row is 1.
2. The coefficients are symmetric with respect to the middle of each row.
3. Each interior coefficient is the sum of the two coefficients above it in the preceding row.

Solved Problem 10.2 Use Pascal's Triangle to determine the coefficients and write the expansion of the following:
(a) $(s+t)^6$ and (b) $(x-y)^5$.

Solution:

(a) $(s+t)^6 = s^6 + 6s^5t + 15s^4t^2 + 20s^3t^3 + 15s^2t^4 + 6st^5 + t^6$

(b) $(x-y)^5$ can be written as $[x+(-y)]^5$ so

$$(x-y)^5 = x^5 + 5x^4(-y) + 10x^3(-y)^2 + 10x^2(-y)^3 + 5x(-y)^4 + (-y)^5$$
$$= x^5 - 5x^4y + 10x^3y^2 - 10x^2y^3 + 5xy^4 - y^5$$

Pascal's Triangle is useful if n is rather small. Its use is not practical for large n. We shall introduce a more practical method for finding the coefficients regardless of the magnitude of n.

We first need the concept of "factorials." A factorial of a number is simply symbolism that represents a particular extended product.

Definition 1: If n is a positive integer,

$$n! = n(n-1)(n-2)\cdots(3)(2)(1)$$

The $n!$ symbol is read "n factorial." It represents the product of all positive integers less than or equal to n.

Definition 2: $0! = 1$.

The $0! = 1$ definition seems arbitrary and illogical, although it will be more apparent subsequently that the definition has merit and is needed for consistency.

Solved Problem 10.3 Evaluate the following: (a) $6!$, (b) $8!$, and (c) $\frac{8!}{5!}$.

Solution:

(a) $6! = 6\cdot5\cdot4\cdot3\cdot2\cdot1 = 720$

(b) $8! = 8\cdot7\cdot6\cdot5\cdot4\cdot3\cdot2\cdot1 = 40,320$

(c) $\frac{8!}{5!} = \frac{8\cdot7\cdot6\cdot5!}{5!} = 8\cdot7\cdot6 = 336$

We now introduce another commonly used symbol.

Definition 3: $\binom{n}{k}$ means $\frac{n!}{k!(n-k)!}$ for $n \geq k$.

Solved Problem 10.4 Evaluate the following: (a) $\binom{6}{4}$, (b) $\binom{12}{8}$, and (c) $\binom{7}{0}$.

Solution:

(a) $\binom{6}{4} = \frac{6!}{4!(6-4)!} = \frac{6 \cdot 5 \cdot 4!}{4!2!} = \frac{6 \cdot 5}{2 \cdot 1} = 15$

(b) $\binom{12}{8} = \frac{12!}{8!(12-8)!} = \frac{12 \cdot 11 \cdot 10 \cdot 9 \cdot 8!}{8!4!} = \frac{12 \cdot 11 \cdot 10 \cdot 9}{4 \cdot 3 \cdot 2 \cdot 1} = 495$

(c) $\binom{7}{0} = \frac{7!}{0!(7-0)!} = \frac{7!}{0!7!} = \frac{1}{1} = 1$

We can now use the factorial symbolism to find the numerical coefficients in the expansion of $(a + b)^n$. These coefficients are called the *binomial coefficients*.

In the expansion of $(a + b)^n$, the term containing $a^{n-k}b^k$ has coefficient:

$$\binom{n}{k} = \frac{n!}{k!(n-k)!}$$

for nonnegative integers n and k and $n \geq k$. Some useful properties of binomial coefficients follow.

For nonnegative integers n and k and $n \geq k$,

1. $\binom{n}{0} = 1$

2. $\binom{n}{n} = 1$

3. $\binom{n}{k} = \left(\frac{n}{n-k}\right)$

We can now state the formula for the expansion of $(a + b)^n$. It is called the *Binomial Theorem*.

If n is a positive integer,

$$(a+b)^n = \sum_{k=0}^{n} a^{n-k} b^k$$

$$= \binom{n}{0} a^n + \binom{n}{1} a^{n-1} b + \binom{n}{2} a^{n-2} b^2 + \binom{n}{3} a^{n-3} b^3 + \cdots$$

$$+ \binom{n}{n-1} a b^{n-1} + \binom{n}{n} b^n$$

Index